MXac
090115

Pyrite

Pyrite

A Natural History
of Fool's Gold

DAVID RICKARD

OXFORD
UNIVERSITY PRESS

OXFORD
UNIVERSITY PRESS

Oxford University Press is a department of the University of
Oxford. It furthers the University's objective of excellence in research,
scholarship, and education by publishing worldwide.

Oxford   New York
Auckland   Cape Town   Dar es Salaam   Hong Kong   Karachi
Kuala Lumpur   Madrid   Melbourne   Mexico City   Nairobi
New Delhi   Shanghai   Taipei   Toronto

With offices in
Argentina   Austria   Brazil   Chile   Czech Republic   France   Greece
Guatemala   Hungary   Italy   Japan   Poland   Portugal   Singapore
South Korea   Switzerland   Thailand   Turkey   Ukraine   Vietnam

Oxford is a registered trademark of Oxford University Press
in the UK and certain other countries.

Published in the United States of America by
Oxford University Press
198 Madison Avenue, New York, NY 10016

© Oxford University Press 2015

Library of Congress Cataloging-in-Publication Data
Rickard, David T. (David Terence), 1943–
Pyrite : a natural history of fool's gold / David Rickard.
pages cm
Includes bibliographical references and index.
ISBN 978-0-19-020367-2 (alk. paper)
1. Pyrites.   2. Pyrites—History.   3. Mineralogy.   I. Title.   II. Title: Fool's gold.
QE390.2.I76R53 2015
549'.32—dc23
2014040022

1 3 5 7 9 8 6 4 2
Printed in the United States of America
on acid-free paper

Contents

List of Figures

Preface

THIS BOOK DERIVES from an academic monograph I wrote in 2012 on *Sulfidic Sediments and Sedimentary Rocks* (Amsterdam: Elsevier). The monograph had 801 pages and over 10,000 references. The monograph was a success in circles where weighty academic tomes have a more important function than as a convenient doorstop, and the idea was mooted that I should write a popular version. The result was this book on pyrite. I realized that although the monograph was the first one ever published on sedimentary sulfides, there was an even greater lack of public awareness about the mineral pyrite, a key mineral in the monograph but, more significantly, the mineral that made the modern world.

This book is the result of this idea. It is aimed at a popular audience and covers extensive areas in both the humanities and sciences. Its purpose is to entertain and inform students at all levels and of all ages interested in broadening and deepening their knowledge of the natural world. Extensive discussions of early drafts of the chapters with reviewers led to the format of the book with a relatively focused text illuminated by a liberal scattering of endnotes with further information for the interested reader. I am aware of Stephen Hawking's editor's admonishment that the readership of *A Brief History of Time* would be halved for every equation included in the text. Equations and computations are relegated to the endnotes.

In general I have used metric units but occasionally stray into US avoirdupois measurements. I have tended to use whichever units seem appropriate in the context. In some cases, I have used formal SI (*Système International*) units; younger readers will be familiar with these but they are generally translated in the text for the benefit of the older generation. One of the problems with the natural world is that it includes extremely small objects such as atoms and extremely large things such as oceans. This means that describing these with any familiar unit, such as meters or tons, leads to extremely long numbers, with row upon row of zeros either

before or after the decimal point. In order to shorten these numbers, the
natural sciences use index notation where the number of zeros is written
in superscript next to 10. So 1 million (1,000,000), for example, is 10^6.
Although this is convenient shorthand, it may not be a familiar one to
some readers, and I have translated the numbers into words. I have also
used CE (Common Era) and BCE (Before Common Era) in place of AD and
BC throughout.

The breadth of this work could not have been successfully achieved
without the help of a number of people. Professor Rob Raiswell of Leeds
University again provided the rock on which many of the early ideas were
washed up. He read all the draft chapters and detailed many of the more
egregious errors that the work originally contained. His enthusiasm and
encouragement were key to sustaining my resolve. Val and Paul Thomas
were the lay readers. Their contributions were invaluable. They ensured
that the contents were understandable, and their comments improved the
text immeasurably. Their wise and incisive comments resulted in whole
chapters being rewritten or rearranged. Three anonymous referees provided extensive comments on drafts of the first five chapters, which were
extremely valuable, not only in providing some direction to the final manuscript but also in their encouragement that assured me that I was on the
right course.

Anna Marie Roos generously made available some of her work on
Martin Lister and directed me to sites for Lister's original writings in
Latin. She pointed out to me that the first use of the epithet *Foole's Gold*
was by Timothy Granger and showed a source for the original poem. Doug
Harper generously shared the findings of his research into the first use of
the term as a synonym for pyrite. Benjamin Hallum is currently working
on Zosimos and provided many useful sources about early alchemy in my
search for the origins of the word *marcasite*. Kurda Saied-al-Berezanchi
patiently showed how the Arabic words for *marcasite* and *pyrite* were pronounced. Lesley Robertson contributed some of her boundless knowledge
of Dutch microbiologists. I consulted original sources wherever possible,
and this put a great burden on the librarians in the Cardiff University
Science Library. I thank particularly Heather Mann and Debbie Burt for
chasing down originals of many esoteric manuscripts.

I was fortunate to meet my future Oxford University Press editor,
Jeremy Lewis, at a conference in the United States. His enthusiasm and
expert direction has been quite exceptional. Jeremy read all the work and
made detailed comments that knocked it into a shape worthy of publishing.

Of course you will no doubt find many errors in the book, both perceived and real, and these are entirely down to me. The real ones are caused mainly by my great age and consequent infirmity of brain; the imaginary ones are caused by a lack of clarity in my text. I apologize in advance for both.

Prologue

WE ARE GOING on a journey, you and I. It will be like no other journey you have ever been on because our guide will be the brassy yellow mineral pyrite, sometimes known as fool's gold.

Pyrite will take us to times and places far into the past and to the furthest corners of our planet. We will see the tangerine skies and chocolate-brown seas of the young Earth thousands of millions of years ago and track pyrite in the oldest rocks on the planet. We will search for the origins of life on Earth, discover how pyrite may have been involved, and, in our grand tour of the countless millions of years of Earth history, see how pyrite has contributed to the evolution of early life, which ultimately produced us.

We will dive to the ocean depths and discover the wondrous oases of life on the ocean floors thriving in the dark, three miles or more beneath the ocean surface. We shall see these isolated communities of strange life forms clustered around smoking volcanic vents that emit black pyritic water at temperatures greater than the melting point of lead. And we will visit active volcanoes and see how pyrite and sulfur are formed and discuss why, until quite recently, most people pictured pyrite forming in volcanoes even in regions where there are no volcanoes.

We currently view the history of the Earth, understand much of its present workings, and predict what will happen to our future environment through the prisms of tiny pyrite grains. Much of what we know, and think we know, about the evolution of the planet is from probing the composition of ancient pyrite crystals. These pyrite grains, which are spread throughout the Earth in time and space, preserve much of the evidence of the nature of earlier Earths, as well as the balances and imbalances of our present system. Pyrite shows us that we can affect the Earth environment through our behavior, but the idea that we control the Earth system is merely human, hubristic self-importance. During our journey we will visit the seething microscopic world of bacteria, and pyrite will show us that these tiny workers actually control the global environment and determine the conditions of the world we live in. Guided by pyrite we will inspect the death zones of the present oceans where poisonous, pyrite-related, hydrogen sulfide is destroying oxygen in the waters and where no animals can live. We will see the havoc wreaked on the land by the acid produced by burning pyrite and choke on the fumes of the pyrite-generated smogs in the atmospheres of our industrialized cities.

We will visit with technologists in their workshops of all ages and see how they used pyrite to create the embryos of many of the great industries that permeate our modern civilization. Pyrite is still a core material in current technological developments, and we will tour laboratories where different sorts of pyrite are being designed to order. We will discover microscopic, raspberry-like pyrite grains and find out that Mother Nature has a particular addiction to making them by the billions every day in soils, sediments, and waters all around us—in fact, whenever and wherever she can.

In all of these systems we will be hunting pyrite. We will catch sight of the mineral at key moments in the entirety of the history of humankind. We will visit with early hominids and see how pyrite was used to master fire, and we will join their family gathering around the hearths to hear the earliest stories and listen to the first songs of humankind. We will admire the first paintings of the teeming wildlife of our hunting domain in charcoal black from the pyrite-sparked fire and the red ocher from pyrite oxidation. We will call on the Quakers of early-19th-century Philadelphia to find out how the term *fool's gold* originated as an epithet for pyrite and how it has permeated cultures worldwide as a metaphor for false values. We will voyage with the old sea dogs like Cartier and Frobisher to discover new lands and see how they used pyrite as a false promise of gold and riches in order to populate their new colonies. We will travel to the ancient Middle East and investigate the wonderful world of the Arab philosopher-scientists and find out why your jeweler insists on calling pyrite *marcasite* and then charges you a premium for it.

Our journey through space and time with pyrite will show us how this mineral has been at the heart of our present understanding of the world around us: how it has been a core material in the development of great sciences and innumerable scientific breakthroughs. We will look deep into crystals of pyrite and see how humankind learned of the existence, arrangements, and behaviors of atoms; how they produce the characteristic forms of crystals; and how this tells us about the fundamental nature of matter. We will probe the wonderful minds of the medieval alchemists and show how trying to fit pyrite into their philosophies caused increasing problems until the Schoolmen gave up all observational science. We will find that many of the early practitioners of the arcane alchemical art were not, however, the charlatans and tricksters of later times but the founders of the science of chemistry. We will see how pyrite contributed to early medicine by providing alternative inorganic components to the medic's cupboard of herbal remedies.

Can pyrite have a purpose? In this book, at least, its purpose is to guide us through the history of the development of our civilization, culture, technology, and understanding of the world around us. It has been 300 years since Johann Friedrich Henckel wrote his great 1,000-page volume on *Pyrite: The Principal Body in the Mineral Kingdom*. Since then the importance of pyrite has been largely overlooked except in the individual, specialist concerns in which it is has played a central role. You will find that the pyrite-guided journey we are taking together in this book will take in many of the highlights of human endeavor. Our final destination will be an appreciation of the role of pyrite as the mineral that made the modern world and, by reflection, a deeper understanding of this world and our place within it.

Pyrite

I

Foole's Gold

So What Do You Do?

This classic opening gambit at the stereotypical drinks party always throws me. I have been a professor at a university for most of my life, so the easiest answer is that I teach. This is true, but it disguises the reality that much of my waking time has been concerned with research. If I admit this, then it becomes necessary to explain what I actually research. One of my pet subjects is pyrite. But if I let on that I research pyrite, my interlocutors look at me as though I am one of those wonderful beings who haunt the bowels of natural history museums as world experts on a rare species of toad. As with toads, most people in the world have heard of pyrite. They know it is a mineral or stone, and most know that it is also called fool's gold, a familiar theme of moral tales and nursery stories. So the idea of someone studying pyrite is not altogether the stuff of IgNobel prizes. Within the time limits imposed by decent conversation I cannot explain that pyrite is the mineral that made the modern world. I cannot refer them to a book about it since there has not been one published about pyrite since 1725.[1]

This book is an attempt to rectify the situation. In it I contend that pyrite has had a disproportionate and hitherto unrecognized influence on developing the world as we know it today. This influence extends from human evolution and culture, through science and industry, to ancient, modern, and future Earth environments and the origins and evolution of early life on the planet.

The book is aimed at making the subject accessible to the general reader. It is not a scientific monograph, since these handle only the science and are really directed at the converted: the high priests of the cathedral of science and technology and their aspirant novices. It is also not

aimed at being a textbook in the conventional sense: textbooks are generally aimed at specific academic courses and ultimately pave the way for
the students to understand the monographs. Scientific monographs and
textbooks tend to explain in detail how things work but generally do not
tell the audience why the subject is important. This book is an attempt to
explain why pyrite is, has been, and will be important to you and the rest
of humankind: why it is the mineral that made the modern world.

The problem with attempting to explain why pyrite is important is that
this subject covers such wide areas of science and humanities. For example, relevant ancient texts were not written in modern English and need to
be translated. It may be that we can never know what the ancients really
meant by a word or phrase, since meanings are related to the contemporary culture and can even change during a generation. Then there is the
name of the mineral. Unless this is accompanied by some basic description, you can never be sure what the ancient author meant by *pyrite*: Pliny
the Elder described it as a type of millstone, for example. And, of course,
other ancient civilizations used alternate names for the mineral. I present
some of the detective work in tracking down the contribution of pyrite
to the development of ancient civilizations and our modern world. All
this means that the early chapters in this book have already conjured up
some interest in a number of disciplines in the humanities as diverse as
Mesopotamian studies and the history of pre-Renaissance science.

In some ways, we are on safer ground when we come to looking at the
contribution pyrite made in prehistory. Here there is no written record,
and we must rely on concrete relics of pyrite use by humans. Luckily
we have this in the archeological remains of ancient hearths and even of
prehistoric travelers who carried bits of pyrite to light fires. These ancient
hearths reflect the origins of culture as family groups clustered around
the fires telling stories and singing songs, passing down their histories to
future generations. And then there is the cave art, where the first artists'
palettes included black charcoal from pyrite-lit fires and the red ochers
from the weathering of pyrite, and the artists' dark underground studios
were illuminated by pyrite-sourced fires. In this context, the study of pyrite
is not only relevant to archeology but also to understanding the evolution
of human culture.

The role of pyrite in fire-lighting is a feature of all ancient civilizations. It led to the development of the modern chemical, pharmacologi-
cal, and armament industries in which pyrite has played—and continues
to play—a vital role. Although crows and the higher apes use tools, no

species other than *Homo sapiens* has created a sequential series of operations by different individuals that produces a new product. Pyrite has been at the root of the earliest human endeavors of this type. The need to produce a pure product from a pyrite source is the basis of the modern chemical industry. And even today, pyrite is still used in the manufacture of sulfuric acid—the most abundantly manufactured chemical on the planet.

Most ancient medicines were based on biological products such as plants. The introduction of nonbiological materials into medicine can be traced back to pyrite. The production of medicines based on pyrite, such as the alums, paralleled the development of the chemical industry and can be shown to be the origin of Big Pharma, the industrial production of medicines.

One of my regrets is not patenting a pyrite-like iron sulfide I originally synthesized. I could see no use for it, until a far more visionary Chinese entrepreneur patented it for use in fireworks. The red colors you saw at the firework display to celebrate the Beijing Olympics of 2004 were partly due to my iron sulfide. It is appropriate that a Chinese entrepreneur should have developed this product, because it was in China that the first fireworks were produced. The black powder used to power the fireworks was derived from pyrite. These fireworks were developed in China through the medieval period until a series of gunpowder-based weapons were produced, including guns, bombs, and shells. Sulfur, which the Chinese produced from pyrite, is a key constituent of gunpowder and acts as both a fuel and an accelerant, enabling the powder to burn so fast that an explosion is produced. Although sulfur in Europe was often derived directly as the native mineral, pyrite-derived sulfur continued to provide a more reliable source: the politics and economics of localized native sulfur deposits contrast markedly with the almost universal distribution and availability of pyrite deposits. Pyrite constitutes the basis of the modern arms industry as various nation-states have sought to protect and expand their interests over the last millennium.

Pyrite has played a significant role in history under the guise of fool's gold, something that promises great value but is intrinsically worthless. In fact, pyrite is not intrinsically worthless, as we have already seen. Even so, the concept of fool's gold has infiltrated our thinking over the last millennium and beyond, and its beguiling falsity continues to feature in youth culture today.

Perhaps not surprisingly, the heroic tales peoples tell each other about the founding of their nations are often legends created to serve later rulers.

"When the legend becomes fact, print the legend," as Maxwell Scott says to Ransom Stoddard in that classic Western film, *The Man Who Shot Liberty Valance*. Several great nations were founded not on great moral principles as later hagiographers would have us believe but on pyrite. Early explorers encouraged settlement by mistakenly or deliberately using pyrite to support claims that fortunes of gold were to be found in the new territories. Frauds based on fool's gold continue to echo down the centuries of heroic exploration.

The year 2014 was designated the International Year of Crystallography by the General Assembly of the United Nations. The reason is that the science of crystallography is little appreciated by the general public or understood by fellow scientists, apart from the crystallographers themselves. And yet this science has won more Nobel Prizes over the last century than any other subdiscipline. Of the 316 Nobel Prizes in science and medicine that have been awarded since 1901, more than 100 have directly involved crystallography. These included Laue for discovering X-rays and the Braggs for showing that they are waves and not particles, Crick and Watson for the structure of DNA, and Hodgkin for her work on vitamin B12 and penicillin, Sanger on insulin, and Perutz on hemoglobin. The golden crystals of pyrite have played a key role in the development of crystallography, ultimately permitting atoms themselves to be counted, imaged, and probed.

Pyrite crystals stand out in nature and the ancients were naturally curious as to how these were formed. Pyrite was different from the run-of-the-mill rock-forming minerals because it decomposed on heating, producing sulfur gases and leaving behind an inert slag. Pyrite did not easily fit into ancient ideas of how rocks and minerals formed, and yet, because of its economic importance as an ore of copper, zinc, lead, gold, silver, arsenic, and sulfur, its formation had to be explained. This search led to the idea that pyrite formation occurred deep within the Earth and the mineral was brought to the surface by volcanoes. In the late 20th century, hot-water vents were discovered populating the deep-ocean floors, which spewed out pyrite at temperatures as high as 400°C. These deep-sea hot-water vents are powered by volcanism and support an amazing diversity of life forms, many of which had never been seen before. The pyritic vents support isolated oases of life up to 4 km down on the ocean floor with biomass concentrations equaling and sometimes exceeding those of the Amazonian rain forest in intensity. The vents act to cool the Earth as a whole. The pyrite they leave behind in the rocks acts as a fossil tracer of the planet's cooling. This helps us understand how the ancient Earth worked and how planetary plate tectonics evolved. It is this process that causes the continual destruction and renewal of the Earth's surface, which is at present unique to this planet in the solar system and helps explain many of its peculiar features, like life itself, for instance.

And yet many occurrences of pyrite in nature do not appear to be related to volcanism at all. Pyrite can be found in soils and sediments throughout the Earth as myriads of microscopic crystals, often giving the mud a distinct bluish tinge. This pyrite is formed by bacteria that remove oxygen from sulfate in the water, producing sulfide that reacts with iron to form pyrite. These bacteria are forming pyrite all around us: they were first isolated from a ditch outside the laboratory. If you scrabble just beneath the surface layers of soils and sediments in oceans and lakes and rivers and puddles, you will find bacteria making pyrite. More than 90% of the pyrite on Earth is formed by biological, specifically microbiological, processes. Indeed, in contrast with historical ideas, pyrite formation can be closely approximated as a biological process, and volcanism plays only a relatively minor role. Sulfide and pyrite provide the primary energy source for bacteria in the vents, and these bacteria provide the primary nutrient source for the strange biota that cluster around these oases of life in the depths of the oceans.

Bacteria are not only integral to the formation of pyrite in nature, but they also catalyze its oxidation and breakdown. This has key consequences for our environment since the abundance of microorganisms involved in pyrite formation and destruction means that vital characteristics of our environment, such as the amount of oxygen in the atmosphere, is related to pyrite formation. The biological connection also means that the global cycles of key elements such as carbon, nitrogen, and phosphorus depend on pyrite; the connection between pyrite and other metals means that the global cycles of key metalliferous elements such as copper, zinc, and lead and many others are also related to pyrite formation and destruction.

This has been the case since the Earth was formed 4.5 billion years ago. Because of the relative resistance of pyrite crystals to heat, pressure, and chemical reaction, pyrite in ancient rocks has enabled us to see what these ancient worlds were like. To a large extent, our understanding about the evolution of the Earth and its surface environment has depended on the results of probing ancient crystals of pyrite.

Sulfur is a key component of biological materials, and the origin of life requires that sulfur is brought together with carbon, nitrogen, phosphorus, oxygen, and hydrogen in self-replicating molecules. In the iron-sulfur world hypothesis, pyrite catalyzes key reactions in the assembly of complex biological molecules necessary for life. Some of the earliest fossils are single-celled microorganisms more than three and a half billion years old that include small crystals of pyrite in their cell walls. And some of the oldest pyrites have isotopic signatures indicating a biological origin for the sulfur. The consequence is that pyrite has been a key component of the Earth system throughout geologic time. The chemical and isotopic compositions of ancient pyrite are used to trace not only the nature of the Earth's environment in the deep past but also how it evolved. Pyrite has been central to our understanding of why the Earth has an oxygen-rich atmosphere, in stark contrast to its sister planets in the solar system, and when and how this atmosphere developed. Pyrite has helped answer the question: What was it like on the Earth in the past?

And the future? Pyrite will continue to have a key role in both the way the surface environment of the planet will develop in the future and in the evolution of our science, technology, and culture. Pyrite is already playing a significant role in frontier areas of science and technology, such as nano-technology and energy conversion. The remarkable chemical and physical properties of this mineral ensure that it will continue to do so. Likewise, the widespread distribution of huge pyrite concentrations throughout both the land areas and the oceans of the Earth will ensure that pyrite remains an important source of the raw materials needed by a future ten billion human beings.

At the same time, the industry of this huge future population will con-tinue to affect the rate of formation and destruction of pyrite on Earth. Human activity is at present enhancing global pyrite oxidation, and this is contributing to the whole surface environment of the planet, including the oceans, becoming more acidic. The increased oxidation of pyrite through human activities, exposing more of the hitherto safely buried pyrite to the air, is producing an increasing number of global environmental prob-lems such as acid mine drainage, acid soils, and poisonous fogs. The cur-rent human contribution to global warming is expanding the areas of the ocean where oxygen is being removed and pyrite formation is increasing: there are now over 400 dead zones in the oceans where macroscopic life cannot exist. All this has potentially dramatic consequences for the future of our species, although the Earth itself will survive.

Foole's Gold

At the present time, the term *fool's gold* is used as an alternative name for pyrite, a glittering golden mineral of iron and sulfur. Pyrite is the most common sulfide mineral on the Earth's surface. This may not mean much in itself, but the bright golden crystals of pyrite contrast markedly with the usually subdued gray, brown, and black hues of common rocks. The gleam of pyrite catches the human eye rather like a silver bottle top catches the eye of the jackdaw. Throughout human existence, pyrite has naturally attracted attention because of its startling contrast to the familiar, everyday rocks around us.

The gold-like appearance of pyrite combined with its startling contrast with everyday rocks has led to some confusion, at least anecdotally, between the two minerals. Since gold has been valued, excessively some may say, for much of historical time, this confusion has created a wealth of tales about pyrite being mistaken for gold. It has also created an industry, especially in medieval times, of charlatans making money by selling pyrite as gold, by claiming that they can change pyrite into gold, or by producing gold from other materials in their laboratories.

The idea behind the term *fool's gold* is that people can be fooled into thinking that pyrite, which is often worthless, was something highly valued. At the same time, *fool's gold* is used in a metaphorical sense to describe anything that is mistakenly highly valued. This metaphorical usage is far older. For example, a translation of the Cursing of Agade,[2] a Sumerian text from around 2000 BCE, reads: "May your gold be bought for the price of silver, may your silver be bought for the price of pyrite and may your copper be bought for the price of lead!"

Theophrastus (c. 372–287 BCE) was a pupil of Aristotle and succeeded him as leader of the Peripatetic School of Philosophy in Athens. He relates an account of gold being seen in cinnabar (mercury sulfide) ores and how these turned out to contain no gold. Theophrastus reports that Kallius, an Athenian metallurgist from the silver mines in Laurion, attempted to get gold from pyritiferous cinnabar ores at Ephesos in Attica.[3] Pliny the Elder (80 CE) writes that the Emperor Caligula ordered gold to be extracted from the bright yellow arsenic sulfide ore, orpiment.[4] The alchemists were using a trick to produce gold from orpiment well into medieval times. In 1556 Agricola not only described clearly how the trick was done but also recommended death to the fraudsters:

No less a fraud, warranting capital punishment, is committed by a third sort of alchemists; these throw into a crucible a small piece of gold or silver hidden in a coal, and after mixing therewith fluxes which have the power of extracting it, pretend to be making gold from orpiment, or silver from tin and like substances.

The first use of the phrase in English was necessarily much later, since it had to await the development of the English language. The fact that the term was not in wide use in other languages and cultures before this period may also suggest that the phrase is specifically English in origin. For example, the classical writers mentioned earlier may have had the idea of something worthless appearing valuable to a fool, but they did not apparently coin any specific phrase equivalent to *fool's gold*.

In 1570, Timothy Granger wrote a poem[5] on *The XXV Orders of Fooles* where he listed twenty-five different fools, including in verse 11 the phrase *fooles gold*:

> *He is a Foole, that getteth his goods wrongfullye,*
> *For his heires after him, wyll spend it vnthriftelye:*
> *This fooles golde is his God, wrongfullye got,*
> *Why y^u foole, thy golde is muk & clay, knowest thou not?*
> *And as the prouerbe doth shew very playne,*
> *A hood for this foole, to kepe him from the rayne.*

Granger emulated Alexander Barclay's 1509 translation of Sebastian Brant's 1495 theological satire *The Ship of Fools* (*Daß Narrenschyff ad Narragoniam* in the original Swabian dialect), which, according to contemporary metrics, was one of the most popular books ever printed. However, this does not mention fool's gold. In late medieval times there were several days set aside each year as *festa fatuorum* or feasts of fools. These included St. Stephen's Day through New Year's Day, Shrove Tuesday (*Mardi Gras*), Halloween, and local saints' days. During these festivals it was customary to mock everything that was revered, and drunkenness and ribaldry reigned. Consequently, catalogues of fools were popular during this period,[6] but Timothy Granger appears to have been the first to list fool's gold specifically.

Different cultures have different variations on the idea of fool's gold. Thus the Chinese idiom is 此地无藏金 which translates as *no gold is buried here*. It relates to a fool who tries to hide his gold but ends up making it more conspicuous by putting a sign above the burial spot. The metaphor refers to anything that draws attention to something we are trying to keep secret. Western cultures, by contrast, tend to have more literal equivalents. Thus the Spanish is *oro de les tontos*, and it is used in the same literal and figurative sense as *fool's gold*. In German the equivalent is *Narrengold*, which is used in exactly the same way as the English *fool's gold* and is a literal translation of the English phrase. The concept of fool's gold continued in the German literature as a moral metaphor into the 18th century.[7]

The original German expression was *Katzenguld*. This is derived from *Ketzerguld*, or false (literally heretical) gold. *Ketzer* is a 14th-century German word that was derived from the Italian *gazzari*.[8] In Swedish, the equivalent is *Kattguld*, which is used in a similar manner to *fool's gold* as something false and worthless. It refers to white mica, which can glisten like gold in the rocks and boulders that litter the Swedish countryside; it was sometimes known as cat-gold or cat-silver in English. It was used by Carl von Linne (1707–1778) and has since become used to encompass pyrite. I always supposed that it referred to the way cats' eyes glistened in the dark, but it is actually a corruption of the German *Ketzerguld*.

By the end of the Elizabethan age, the concept of fool's gold—in the sense of some ephemeral, worthless treasure—appears to have been well established in English. The proverb "all that glisters is not gold" is a derivative of a Shakespeare line in the *Merchant of Venice* written between 1596 and 1598:[9]

> *All that glisters is not gold;*
> *Often have you heard that told:*
> *Many a man his life hath sold*
> *But my outside to behold:*
> *Gilded tombs do worms enfold.*
> *Had you been as wise as bold,*
> *Young in limbs, in judgement old*
> *Your answer had not been inscroll'd*
> *Fare you well, your suit is cold.*

It appears to have originated in the Latin saying *Non omne quod nitet aurum est* (Not all that shines is gold). The phrase may have originated in the 12th century from a Latin translation of Aesop's fables—although the connection with Aesop's fables seems somewhat tenuous to me. Various versions in English were used by Chaucer (c. 1380 CE). However, it seems that none of these earlier authors had related fool's gold specifically to any particular mineral.

Fool's Gold as Pyrite

The earliest published use of the term *fool's gold* as an alternative description of pyrite may be in the first edition of *The Friend*, a Quaker journal published in Philadelphia in 1829. In an article about the gold mines of North Carolina the author, simply listed as Werner,[10] writes that "The common pyrites, which has gained the expressive name of fool's gold, is the material with which these alchymists generally play."

North Carolina was the only state producing gold in the United States from 1803 to 1828, following the discovery of gold there in 1799 at the Reed mine. It continued as the leading gold producer until gold was discovered in California in 1848. Although started as early placer mining, by the time Werner had written his article most of the mining was underground for lode gold. This is significant since fool's gold is usually associated with hard rocks rather than soils or stream sediments, since pyrite rapidly oxidizes in water. The bright golden color that the unwary might mistake for gold requires a freshly broken rock surface.

It is probable, though not ineluctable, that in order for the term *fool's gold* to be transferred from a purely metaphorical epithet to a synonym for a mineral, gold must have been known locally. There is little gold in the British Isles, though it is not unknown. Thus it is likely that the term was first used in the English-speaking colonies. The North Carolina gold was amongst the earliest found in the English-speaking hegemony, and it is not improbable that the term was first used there. Certainly the writer in *The Friend* in 1829 was not originating the phrase. He writes that pyrite "has gained the expressive name of fool's gold." The suggestion is that the local miners already used the term. The North Carolina gold miners of the early 19th century were largely German immigrants intermixed with Cornishmen. Splendid pyrite crystals are common in the Cornish mines, as well as in the mines of Freiberg. Juxtapose this with the fact that North Carolina produced some of the biggest gold nuggets in the United States[11] at that time, and perhaps a reason for the miners using a put-down for pyrite is understandable. The Germans had their term *Narrengold*, and this was well established in German moral and theological literature. The Cornishmen were also well steeped in the metaphorical idea of fool's gold.

The author of the article in *The Friend*, however, refers to "alchymists" as using fool's gold. This might suggest that the use of fool's gold as a synonym for pyrite was merely a way of explaining the trickery of alchemists.

One of their constant quests was the philosopher's stone, which turns base metals and other materials into gold. Demonstrations of these transmutations were popular throughout the Middle Ages up until the scientific renaissance of the 18th century. In French, the literal translation of fool's gold is *l'or des fous* and it has a similar meaning to the English *fool's gold*. However, there is an additional fable related to *l'or des fous* about some villagers in southeastern France who collected pyrite and heaped it in piles outside, hoping that sunshine would turn the mineral into gold.[12] The story seems to be a satire on the alchemists turning base metals into gold. The thread of false values still runs through this tale.

This French tale is an exception, however. Prior to the 19th century, the use of the term *fool's gold* to describe a real material referred to the false gold produced by the tricksters and not necessarily to pyrite. Miners, rather than alchemists, first applied the term to pyrite. It was in the local vernacular by the time it was recorded in 1829. Finally, it is worth noting that the use of the term *fool's gold* as a synonym for pyrite was originally written up in a religious journal published by the Society of Friends. This background is significant, since the Quakers would not only have been well aware of the moral and religious hinterland of the term *fool's gold* but it would have been a basic tenet of their *Weltbild* or world image.

There is further evidence for this dating of the original use of the term *fool's gold* as a synonym for pyrite. Henckel does not mention fool's gold in this great volume on pyrite published in 1725, nor does it figure in later English translations. Wallerius does not mention fool's gold in his mineralogic textbook of 1747,[13] nor does it occur in the later German and French translations. Both Henckel and Wallerius wrote in the vernacular rather than, as Agricola in 1556, in Latin. It is more probable that these authors would have mentioned the term if it were common amongst the mining communities of Germany and Sweden with which they were embedded.

The initial introduction of the term *fool's gold* for pyrite in 1829 introduced a welter of similar published accounts, all in the United States. In 1834, Thomas R. Gordon wrote:[14]

Jenny Jump in the N.W., a gold mine, is said to exist. Preparations have ostensibly been made for smelting the ore, but the "wise ones" have little confidence in the undertaking, and consider the minerals discovered, if any, to be pyrites or fool's gold.

The term had obviously spread to New Jersey, although this publication was also printed in Philadelphia, Pennsylvania, like the original reference in *The Friend* of 1829.

By 1840, the term had been included in an introductory geology textbook.[15] Edward Hitchcock, the state geologist of Massachusetts who was to become the third president of Amherst College, wrote that "By no mineral substance have men been more deceived, than by iron pyrites; which is very appropriately denominated fool's gold."

By 1845, the term had spread, through Massachusetts and to Vermont. In the *First Annual Report of the Geology of the State of Vermont*, the Vermont State Geologist Charles Baker Adams reported on the "money bed" found in the eastern part of Brandon. A mysterious ore was dug up and locked in a small building, which was actually iron pyrites (fool's gold). Adams noted[16] that the "money bed" was appropriately named since "the money lost here would have been a handsome addition to the funds of a Geological survey." Adams had studied at Amherst College, where Hitchcock taught, and lectured there until 1838 to move to Vermont.

James Dwight Dana, the preeminent 19th-century American mineralogist, did not mention fool's gold in his 1837 *A System of Mineralogy*.[17] However, it also does not appear in later editions of this monumental work, and this may simply reflect Dana's more strait-laced approach to mineralogy. My secondhand edition of Rutley's *Elements of Mineralogy*,[18] a standard textbook of all UK geology students in the 1960s, also does not refer to fool's gold. This is significant, since it includes many of the other old miner's terms for minerals, including *capillary pyrites* for example. It may suggest that the use of the term *fool's gold* for pyrite was from the vernacular of North America rather than the British Isles.

In 1882 the term *fool's gold* appeared in an article in the *Boston Journal of Chemistry and Popular Science*. This article has a Mark Twainish ring about it, and the epithet "Fool's Gold" appears to reflect the popular style in American journalism at the time. The anonymous author notes: "Scarcely a day passes when chemists, mineralogists and editors do not receive from some deluded individual a specimen of some yellow stuff to analyze."

Again the "deluded individual" is very much in the prevailing American Zeitgeist. He goes on to state that "nine times out of ten, at least, this yellow material is a sulphide of iron, to which, for this very reason, the name has been given that heads this article,—fool's gold. In scientific works it is called iron pyrites."

The article may have been written by Williams James Rolfe, the editor of the journal through its various metamorphoses from 1868 to 1910. Rolfe was unusual by comparison with modern editors of chemistry journals since he was also one of the leading Shakespearian scholars of the time: he edited nearly all of Shakespeare's plays and wrote three books on Shakespeare. Rolfe also published volumes on English literature, history, and physics and edited the works of classical authors. It seems likely that *fool's gold* is the sort of phrase he would enjoy.

In 1899 Edwin Tappan Adney[19] mentioned fool's gold in a footnote about the problem of identifying gold in the Klondike: "Fool's gold, so often mistaken by the inexperienced, is sulphuret of iron, or iron pyrites."

Adney was a photographer and artist who was born in Athens, Ohio, but traveled widely in the Yukon, ultimately becoming a Canadian citizen. Fool's gold had reached the furthest parts of North America by the end of the 19th century.

Fool's gold may have originated as a metaphor in Europe, but it seems that it developed as a synonym for pyrite in the 19th-century United States. It initially spread from the gold workings in North Carolina through the eastern states of the United States partly through students at US colleges and partly through the coincidence of the location of journal and book publishers in key US cities.

Pyrite and the Founding of Canada

During the 16th century, Spain was shipping huge amounts of gold and silver back to Europe from its American colonies in Peru and Mexico. England and France looked on with envy, and it became state policy in both countries to find alternate sources of gold in the Americas.

In 1534 the French King Francis I commissioned the Breton navigator Jacques Cartier to "discover certain islands and lands where it is said that a great quantity of gold and other precious things are to be found." Although Cartier's first voyage only probed the mouth of the St. Lawrence River, the results were sufficiently exciting to warrant a second voyage.[20] In 1536 Cartier penetrated the St. Lawrence River as far as what is now Quebec City. There he met an Iroquois chief who told him of the legendary Kingdom of Saguenay to the north, where diamonds and gold were abundant. The Saguenay River is shown in

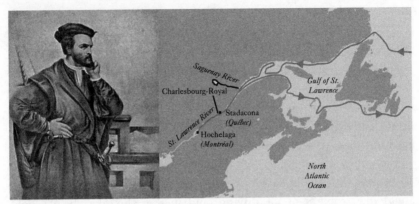

FIGURE 1.1. Jacques Cartier and the route of his second voyage in 1536. The Native American village names are shown together with their modern equivalents; Charlesbourg-Royal was founded by Cartier on his third voyage in 1542 but was subsequently abandoned. This image of Jacques Cartier is from the St. Malo portrait painted in 1839, almost three centuries after his death. There is no contemporary likeness of Cartier but this is the accepted (suitably heroic) image.

Figure 1.1 and became an important trade route into the interior. Cartier shipped the chief back to France, where he continued to expound on the riches of Saguenay. The chief's tales confirmed the king's own beliefs, and Cartier was commissioned to return on a third voyage with a larger fleet, including colonists, mainly convicts.

Cartier returned in 1542 and founded the settlement of Charlesbourg-Royal on the banks of the St. Lawrence. Although Charlesbourg-Royal was abandoned in 1643, Cartier had introduced the name *Canada* for the lands along the St. Lawrence, and this became the designation for the later French colonies.[21]

Cartier's crew found large quantities of "diamonds" and "gold" in the area, which they shipped back to France. They turned out to be quartz crystals and pyrite. The French coined a phrase *faux comme les diamants du Canada* (as false as Canadian diamonds)[22] referring to any false treasure as a result of this exploit. They apparently did not latch on to the idea of the pyrite being "fool's gold."

Frobisher's Scam

The English must have known about the French explorations and set about mounting their own expeditions to North America. During the reign of Elizabeth I, the only gold the English could get cheaply was by stealing it from the Spanish treasure fleets. This activity created an upsurge in highly skilled pirates, sometimes working with letters of marque and calling themselves privateers. One such entrepreneur was Martin Frobisher, one of the sons of the Master of the Royal Mint.

Frobisher was arrested four times for piracy but was freed each time with a warning not to do it again. He noted that the Spanish gold came from new overseas colonies and came up with a splendid scheme to discover a new route to China and the Indies via the north of Canada: the Northwest Passage. Frobisher's subsequent activities confirm that he and his financial backers were not exclusively concerned with the discovery of new trade routes but more interested in finding gold. The idea was reasonable: if the Spanish found gold in the New World, it would be possible for the English also to find gold there. Unfortunately, the Native Americans of eastern North America, unlike their southern compatriots, had little knowledge of metals, let alone gold and silver.

Frobisher's persistence finally resulted in financial support from the Muscovy Company, especially from Michael Lok, the company's director. Lok arranged the money for two 25-ton barks, a 10-ton pinnace, and a crew of thirty-five. One of the barks and the pinnace were lost on the voyage, and five of the crew were captured by Inuits in Labrador and never seen again. Frobisher established mines on the remote Kodlunarn Island, which he named the Countess of Warwick's Island after one of the expedition's investors. The remoteness of this location is apparent in Figure 1.2. The remains of his mines, some buildings, and a slipway for the ships and the ore have been preserved in this sub-Arctic environment.

Frobisher originally brought back a piece of "black stone." It seems that Frobisher first thought that this was sea coal that could provide a useful source of fuel for later settlers. However, it somehow fell into the hands of the wife of one of the adventurers,[23] who put it in the fire. All this seems reasonable if it was thought to be sea coal. But then, for some unknown reason, she quenched the stone with vinegar and it developed a sheen of "bright marquesset of gold." This unlikely tale was the origin of the idea that Frobisher had found gold. Lok gave pieces of this black stone to the crown assayist, a Mr. Williams of the Tower of London, who reported no gold and that the stone contained *marquesset*, a term that we might interpret today as pyrite. Indeed, bright golden pyrite crystals are commonly found in coal. Williams suggested that a gold refiner, a certain Mr. Wheeler,[24] might check the result, and he confirmed the lack of gold.

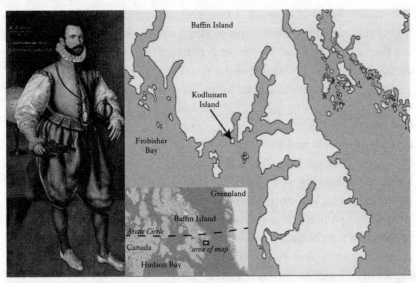

FIGURE I.2. Sir Martin Frobisher and the location of his mines for fool's gold on Kodlunarn Island in the remote Canadian north. The image is from a painting by Cornelius Ketel in 1577, now in the Bodleian Library.

A third assayist, George Needam, also reported that the stone contained no gold. He then gave a small piece to Jean Baptiste Agnelli, who produced a small pile of gold from the sample within three days. Lok gave Agnelli a further sample of the black stone, and again he returned a small quantity of gold. When Lok queried him about why his results differed from those of the other three assayists, Agnelli replied that it was because he knew how to flatter nature.

The results were enough for Lok and Frobisher to seek funding for a second expedition. The Cathays Company stumped up the funds this time, and three ships set sail in 1577. The real purpose of the expedition could not be questioned: amongst the 150 crew were "miners" and "refiners." Like many of Frobisher's exploits, the "miners" and "refiners" were a bit of a sham. They were later described as "showmakers, taylores and other artificiers" since the original miners were "intimidated by the commissioners" and withdrew.[25]

Frobisher was actually directed to postpone matters of exploration to another time and concentrate on the gold prospecting. This time he brought back 140 tons of rock claiming it was gold ore. The ore was taken to Admiral Sir William Wynter's house, Lydney Manor, in the Forest of Dean, just 35 miles northeast of where I am writing. At Lydney, a German metallurgist (and assistant to Agnelli), Jonas Schutz, built a furnace to smelt the ore. This seems possible since the Forest of Dean was a major iron mining and smelting district at the time. Schutz produced gold worth £40 from a hundredweight of the rock. However, goldsmiths and assayists of London could find no trace of gold in the ore. At this point the fortune teller and healer Burchard Kranich, or Dr. Burcot as he styled himself in London, was brought in and confirmed Schutz's results.

This time the queen was taken in by the scam. She presented Frobisher with a gold chain and the wherewithal for an expedition of fifteen ships with a view to setting up a colony and mining the ore. He returned with a further 1,100 tons of ore, and a special smelter was set up to process it. However, no gold was ever produced, and, finally, the scam was discovered.

We are fortunate in having detailed reports of all Frobisher's voyages by members of his expeditions,[26] as well as detailed records of the extensive litigious activities that followed these voyages. It is clear that the expedition reports *marquesset* in the rocks Frobisher collected. At the same time the reports quote variations on the proverb that all that glistens is not gold, so his companions appear to have been fully aware of the situation. Burchard Kranich seems to have been part of the scam. Kranich was credited with curing Queen Elizabeth of smallpox. Thus his reputation in the royal household was *sans pareil*. Indeed, she wanted to give him a grant and was stopped only by Lord Burghley. So when Kranich verified the gold content of Frobisher's ore the queen was, to say the least, predisposed to believe him, even against the results of more conventional assayists. Kranich used a special flux to refine the ore. However, a sample of this flux was stolen and analyzed separately. It turned out to be rich in precious metals. In effect, Kranich was salting the ore.

So was this the source of the phrase *fool's gold* as a synonym for pyrite? There is no contemporary evidence for the written usage of the phrase, and I rather think not. It seems to have been part of a series of ongoing scams at that time to sucker a society that was desperate for gold. However, it probably added to the general mythology of pyrite. The importance of these fool's gold discoveries is that they powered a series of expeditions to the eastern coast of North America and added to the knowledge about the technology required for the voyage across the Atlantic Ocean as well as the coastal geography.

Pyrite and the Founding of the
United States of America

The Frobisher affair rumbled on through the courts until 1620 when Michael Lok died. Thus it was still a *cause célèbre* in London when the Virginia Company of London was formed in 1606 with the express purpose of finding gold in North America.

The Virginia Company rented two ships, bought a pinnace, and landed in Chesapeake Bay in 1607. The expedition founded the first permanent European settlement at Jamestown on Chesapeake Bay and this would become the birthplace of the United States of America. It could be argued that the United States was founded on fool's gold.

The colonists were driven by promises of gold, and John Smith, one of the original council leaders and the third elected president of the colony, was under constant pressure to find gold. John Smith's problem was that he was not the sole leader of the first group of colonists. Indeed, at sea Christopher Newport, the captain of the *Susan Constant*, the ship that brought the settlers, was the absolute ruler. He was the senior captain of the largest vessel. And he was convinced that gold was to be found in the area. We can see that John Smith's position was delicate. The settlers included the crew members who were in thrall to Christopher Newport rather than John Smith. And even the nonsailors would have spent the crossing under Newport's command. So John Smith was forced to accept Captain Newport's views.

In June 1607 they had found a "native American mine" of clay sand with sparkling golden crystals. This was not a mine but merely a sedimentary deposit, and the description of it as a mine was probably an added public relations spin. Christopher Newport carried some 1,100 tons of this ore back to England in the *Susan Constant* for assay. The golden crystals turned out to be pyrite. We know that it was pyrite since the same deposits can be visited today.

By contrast with the Frobisher exploit, which was still reverberating through the courts in England, it seems that there was never any doubt about the nature of the ore when it arrived. It is said that Smith was originally doubtful that the mineral was gold, but he bought some time in the internal politics of the colony with the shipment. However, Captain Newport was to make several return voyages across the Atlantic with more settlers initially funded by the false promise of the fool's gold.

The discovery of metaphorical gold—in the sense of a valuable commodity—had to wait until 1612, when John Rolfe started growing tobacco and the sale of the first Virginia tobacco in England in 1614 sparked a major migration.

As a codicil to these events, the Frobisher and Smith actions in sending pyrite back to England as gold ultimately led to the execution of the man who popularized tobacco in London: Sir Walter Raleigh. Raleigh obtained his half-brother's, Sir Humphrey Gilbert's, charter from Queen Elizabeth I to establish a colony in North America. He never made the journey himself but delegated it to Phillip Armadas and Arthur Barlowe. The driving force in this case was the idea of establishing a base from which to raid the gold in the Spanish treasure fleets. Raleigh funded the expeditions himself. Armadas and Barlowe landed on Roanoke Island in what is now North Carolina in 1584. They returned to England, and Raleigh sent a larger fleet under the command of his cousin, Sir Richard Grenville. Ralph Lane, who was delegated to operate the charter together with Grenville, was left on Roanoke Island with 107 men in 1585. Interestingly from the point of view of this narrative, one of the settlers was Joachim Gans, a distinguished Bohemian metallurgist who was later rescued from Roanoke by Sir Francis Drake. Archeologists have found lumps of smelted copper and a goldsmith's crucible at the Roanoke site, probably attributable to Gans's work. So perhaps the idea of gold prospecting was not entirely absent from the original quest of these early adventurers. A new fleet sent out by Raleigh in 1587 found little trace of the original colony, but another 150 settlers were left on Roanoke Island. The new colony's governor left for England to get help in 1587, leaving 115 colonists behind. The coming of the Spanish Armada delayed the necessary assistance being organized, and it was 1590 before he arrived back in Roanoke. He found the colony deserted, and the mystery of what happened to these Roanoke settlers remains to this day.

Raleigh led expeditions to South America in 1595 and 1618 to find gold in the Spanish domains. He claimed to have found El Dorado up the River Orinoco. He was executed in 1618 ostensibly because he had fought the Spaniards against the king's express orders and due to the application of the terms of an earlier death sentence for treason in 1603. In reality, James did not believe Raleigh and thought that his claims were a fraud.[27] You can see how the effects of the Frobisher, Smith, and Cartier affairs affected James's thinking. The sting in the tail is that gold did indeed exist in the Orinoco mines, and Raleigh was telling the truth all along.

Jamestown was founded on the deliberately propagated belief that gold was to be found in those areas, whereas this turned out to be worthless pyrite. Even so, the establishment of Jamestown ultimately led to the tobacco plantations, the first legislature, and the establishment of the Virginia colony. Later, in 1620, the *Mayflower* settlers set sail from Plymouth, England, financed by the London Company, a rival group of investors to the Virginia Company who financed the Jamestown expedition, to establish a settlement on the East Coast of America. They founded the Plymouth settlement in present-day Massachusetts. This later settlement was facilitated by the success of the Virginia Company's investment in Jamestown. It heralded the mass immigration to North America by European immigrants and ultimately led to the foundation of the United States of America.

The effects of the original Jamestown settlement on the history and development of the United States continued for some centuries. Many people are surprised, for example, to learn that the Pilgrim fathers were later immigrants when Jamestown already had a legislature and the colony of Virginia was well on its way to being established. After the Civil War the northern, Massachusetts-based Plymouth settlement was written up by the victorious Union hagiographers, and these later settlers became the founding fathers of modern legend.

It is interesting that the first European settlement was established on a search for gold and that this turned out literally to be fool's gold or pyrite. The United States would ultimately have been settled anyway, but the actuality is that the origins of that great nation were originally based on pyrite.

Fool's Gold in the Modern World

The idea of the old prospector pushing open the swing doors of the small-town bar, shouting that he had made it rich and trying to pay for his whisky with fool's gold is probably a Hollywood myth. No experienced prospector would mistake pyrite for gold. Even James William Marshall, the sawmill builder who first found gold in California in 1848, knew enough about the metal to know that it is malleable. He took the yellow mineral he had noticed in the sawmill tailrace and simply hit it with a hammer. If it was pyrite, it would shatter but, being gold, it flattened.

By contrast, less experienced people, like the tenderfoots of the gold rush days, may not realize the difference between pyrite and gold. I was chair of a university department that had a 25-year lease on a gold mine in west Wales. The queen actually owned the gold, as she does all precious metal deposits in the United Kingdom, and we leased the mine from the Crown Commissioners. We used the mine as a training base but had permission to set it up as a tourist center in the summer months. We expanded the underground visit trail by making a new adit that penetrated a rich vein of pyrite. The students piled the pyrite up in the middle of the mine yard for tourists to help themselves to. One student told me he persuaded his milkman that it was Welsh gold and he paid his milk bill with a jam jar of it. The fool's gold had some value after all.

There is a twist in the tale. The Romans worked our mine for gold, which is a classic type of gold deposit where the gold is closely associated with the pyrite. The Romans collected the free gold by simply crushing the rock and washing out the lighter minerals. In fact much of the gold is contained within the pyrite itself in this type of ore and can be chemically extracted. So, as discussed in Chapter 10, fool's gold may not be so foolish after all.

I regularly went to the local Odeon cinema on Saturday mornings with hundreds of other children for the film show which we called *the pictures*. We were shown two films, usually a cartoon and a Western cowboy adventure for sixpence, equivalent at least officially to five pence or perhaps a dime nowadays. One of our heroes was the big-hatted, black-suited, cowboy with the pair of silver, white-handled six-shooters and the white horse, Hopalong Cassidy, played by William Boyd. The Hopalong Cassidy films were the most successful B-Western series ever made and continued over many years from 1935 through to the 1950s. The good guy always won, and the continuing moral was that of fair play. In 1946 Boyd made a film called *Fool's Gold*, an oater that was based on the idea that the baddies were intending to substitute gold-plated copper bars for real gold ingots while stealing the real ones.

Hopalong Cassidy's *Fool's Gold* was not the first film with this name. The earliest was probably the 1916 silent film by Edward T. Lowe, made in Chicago with Nell Craig and Darwin Carr. This film has nothing to do with mining, and the title refers to worthless money that can result from gambling. However, it demonstrates that the epithet *fool's gold*, used in this metaphorical sense, rapidly penetrated this new popular culture. This was reinforced in the film industry in 1919 by another film with the same title, although this one did have a mine in it (actually two mines); even so, the title refers to the moral that human relationships are more important than gold. The most recent film with this title is the 2008 romantic comedy with Kate Hudson and Mathew McConaughey about a couple hunting buried treasure. This closely reflects one of the earliest uses of the term *fool's gold* in *The Atlanta Constitution* in 1888.[28]

The use of *fool's gold* in modern culture is well established. I cannot find a direct reference in modern poetry to compare with Timothy Granger's first use of the term; however, it appears frequently in popular music and in song titles. In 1989, *Fool's Gold* became the most famous song of *The Stone Roses*, a world-renowned British rock band. It was basically a dance tune that went on to be sampled and remixed by several artists and used in numerous films and computer games. The lyrics are weird, and it is unclear what the fool's gold refers to—although they appear to imply that gold or wealth is worthless and weighs you down, and that might be the tag reference.

Another world-famous British group, Procul Harum, wrote a song in 1975 called *Fool's Gold*, which included the lyrics

> *I was trying hard to win*
> *save the world and be the king*
> *I was out there in the race*
> *trying hard to force the pace*
> *Fool's gold fooled me too*
> *bright and shiny looked brand new*
> *Fool's gold broke my heart*
> *shone so brightly then fell apart.*

This compares favorably, I suppose, with the doggerel of Timothy Granger. Keith Reid, one of the song's writers, referred frequently to gold in many of his songs.

The rhythm-and-blues singer Frank Ocean also wrote a piece called *Pyrite (Fool's Gold)* in the chorus of which he refers to

> *But even a fool knows when it's gold, gold, gold*
> *I know pyrite from 24 karat*
> *Cubics from genuine diamond (yeah)*

The writer used a quite literal interpretation of fool's gold as a means for telling genuine from false romantic feelings.

Fool's gold appears prevalent in modern popular culture as a vehicle for expressing teenage angst and bewailing the falseness of some adult values as well as the romantic perfidy of partners. The expression may have reached its apogee in the reaction to the worldwide financial crash of 2008, which is often considered to be the worst financial crash since the Great Depression of the 1930s. It was based on the false premise of a permanently rising value of the housing market, and the fool's gold metaphor perfectly encapsulates this naïve belief. It was enhanced by reckless gambling by finance institutions in the trading of derivatives: financial vehicles that have no intrinsic value but are merely varieties of bets on the values of real assets. The *Financial Times* journalist Gillian Tett published a best-seller titled *Fool's Gold* in 2009; it was subtitled *How Unrestrained Greed Corrupted a Dream, Shattered Global Markets and Unleashed a Catastrophe*. In this the author, originally trained as a social anthropologist, relates how a group of bankers ("a small tribe") at J.P. Morgan built a monster that got out of control and helped destroy much of their industry. This group traded in derivatives, which were based on housing loans. They essentially acted as bookmakers, spreading risks to other banks in a win-win situation—as long as the values of the primary assets (the houses) did not all collapse. The scam has all the characteristics we associate with the idea of fool's gold: a worthless asset believed by some people to be of real value.

The financial crisis continues today. J.P. Morgan announced a loss of US$2 billion on derivative trading in May 2012 that had risen to US$9 billion by June. To put this in perspective, they may have some US$70 trillion tied up in derivatives. This has given rise to the concept that these financial institutions are too big to fail, since the value of their debt is larger than the global economy. The idea of a false certainty that is expressed in the term *fool's gold* seems to be very appropriate in our current world.

Notes

1. J.F. Henckel. 1725. *Pyritologia, oder: Kieß-Historie, als des vornehmsten Minerals, nach dessen Nahmen, Arten, Lagerstätten, Ursprung* (Leipzig: J. Chr. Martin), 1,008pp. But also see D. Rickard. 2012. *Sulfidic Sediments and Sedimentary Rocks* (Amsterdam: Elsevier), 801pp., which contains much about pyrite, since this is the principal mineral in sulfidic sediments and sedimentary rocks. It also lists about 10,000 references if you need to look up a primary source about anything in this volume.

2. Faculty of Oriental Studies. 2006. ETCSL Project (Oxford: University of Oxford). The translation of *za-ha-am* on line 243 as *pyrite* is surrounded with uncertainty. It literally means *shining stone*, and the translator might have been infected by the fool's gold tradition.

3. E.C. Caley and J.F.C. Richards. 1956. *Theophrastus on Stones* (Ohio State University Graduate School Monograph).

4. J. Bostock and H.T. Riley. 1855. *Pliny the Elder, The Natural History* (London: Taylor and Francis). G. Agricola. 1556. *De Re Metallica* (Baseleae: Froben). Translated by H.C. Hoover and L.H. Hoover. 1950 (New York: Dover).

5. T. Granger. 1570. *The XXV Orders of Fooles* (London: Henry Kykham). I thank Dr. Anna Marie Roos, Oxford University, for bringing this to my attention.

6. For example, the catalogue of fools in R. Copland's *Jyl of Braintfords Testament*. 1567 (London: Copeland).

7. For example, J.G. von Kaysersberg's *Der Sittliche Narrenspiegel*. 1708.

8. The Latin *cathari*, which in turn gave the name to the Cathars, the great Christian heresy of southern France in the 12th and 14th centuries.

9. Act II, Scene VI, Prince of Morocco.

10. R. Smith, ed. *The Friend*. 1829. Vol. 1 (Philadelphia, PA: John Richardson), p. 92.

11. J.D. Dana. 1837. *A System of Mineralogy: Comprising The Most Recent Discoveries* (New York: Wiley & Putnam) reports a single gold nugget from Cabarra County, North Carolina, weighing "twenty eight pounds avoirdupois," worth around US$700,000 today.

12. The connection between gold and the Sun has a history extending back over 4,000 years to the original proto-Indo-Europeans. It was revitalized in the 16th century by the discovery by the Spanish that the native peoples of Central and South America possessed large gold deposits. Indeed, the Incas had a helio-centric religion where gold represented the Sun. This may have affected subsequent Spanish thought. They may have given up their possessions on the west coast of the present United States and Canada somewhat less easily than would have been the case if they had not had an atavistic feeling about a connection between the Sun and gold. Of course, gold was found in these less sunny climes less than fifty years after the Spanish left these regions.

13. J.K. Wallerius. 1747. *Mineralogia eller Mineral-riket* (Stockholm: Uplagd på Lars Salvius). Published in German (1750) and French (1753).

14. T.R. Gordon. 1834. *A Gazetteer of the State of New Jersey* (Philadelphia, PA: Daniel Fenton), p. 162.

15. E. Hitchcock. 1840. *Elementary Geology* (New York: Dayton and Saxton), p. 62. This textbook was to run to thirty-one editions.

16. C.B. Adams. 1845. *Annual Report of the Geological Survey of Vermont*. Vol. 1 (Burlington, VT: Chauncey Goodrich), p. 31.

17. Dana. 1837. *A System of Mineralogy*, 633pp. Mind you, fool's gold is not mentioned in the 8th edition from 1997 either.

18. F. Rutley. 1874. *Elements of Mineralogy* (London: Thomas Murby). This textbook went through twenty-seven editions, and the updates were continued after Rutley's death in 1904 by H.H. Read through to 1965 and C.D. Gribble in 1988. Both Rutley and Read were professors at Imperial College, London, and this was naturally the Imperial College students' basic mineralogy textbook. My edition was from 1936.

19. T.E. Adney. 1899. *The Klondike Stampede* (New York: Harper).

20. For details of Cartier's voyages see, e.g., J.F. Blashfield. 2002. *Cartier: Jacques Cartier in Search of the Northwest Passage* (Minneapolis, MN: Compass Point Books).

21. The word *Canada* derives from the Huron-Iroquoi word for *kanata*, meaning village. It is a familiar problem for cartographers working in areas where they do not speak the language. There is said to be a South Pacific Island called "I don't know" in the local language.

22. In fact, Canada became a significant diamond producer at the beginning of the present millennium with discoveries in the Northwest Territories. It is currently the world's third largest producer.

23. Some say this was Frobisher's wife.

24. Wheeler is only known through Michael Lok's evidence: "And thereupon I gave an other small pece to one Wheler gold fyner by Mr Williams order. He aunswered also that he made proof and founde it but a marquesite stone." See V. Stefansson. 1938. *The Three Voyages of Martin Frobisher*. Vol. 2 (London: Argonaut Press), p. 83.

25. T.A. Rickard. 1947. *The Romance of Mining* (Toronto: Macmillan).

26. These accounts are collected in V. Stefansson. 1938. *The Three Voyages of Martin Frobisher*. They include Dionyse Settle's 1577 account of the second voyage on which he served as "a gentleman," Thomas Ellis's (described as a sailor and one of the company) and Edward Sellman's (Michael Lok's representative) accounts of the third voyage, and George Best's 1578 account of all three voyages. George Best was a lieutenant on the second voyage and captained one of the ships in the third voyage. See also G. Beaudoin and R. Auger. 2004. Implications of the mineralogy and chemical composition of lead beads from Frobisher's assay site, Kodlunarn Island, Canada: Prelude to Bre-X? *Canadian Journal of Earth Sciences*, 41:669–681. D.D. Hogarth and J. Loop. 1986. Precious metals in Martin Frobisher's "black ores" from Frobisher Bay, Northwest Territories. *Canadian Mineralogist*, 24:259–263.

27. P.R. Sellin and D. Carlisle. 2011. Assays of Sir Walter Raleigh's ores from Guyana, 1595–96. *The Ben Jonson Journal*, 18:274–286.

28. "Fool's Gold" was used in a headline in *The Atlanta Constitution* in 1888, the daily newspaper of that southern US city. The phrase was used to describe the search for a pirate hoard.

2

Pyrite and the Origins of Civilization

PYRITE IS AN often-overlooked material today although it has been instrumental in enabling many aspects of our modern culture and industry. This bright, brassy mineral is the most abundant metal sulfide in the Earth's crust and provides a marked chemical contrast to the duller silicates and oxides that constitute most rocks. Most people today are familiar with the mineral, even though they do not know its details, because it stands out in the natural environment and because of the connection with fool's gold.

Pyrite has been a source of both metals and sulfur since ancient times, and both of these commodities have been key to our civilization. The mineral is easily decomposed by heat with the production of sulfur, sulfur oxide gases, and a metal-rich slag. It oxidizes readily in aerated water to form red and yellow ochers that may be used as pigments. It commonly occurs with other valuable metals that may be extracted by leaching or heating with various fluxes. In summary, it is an exceptional mineral whose benefits were readily available to primitive societies and have led to the development of our modern civilization.

The Gods of Fire

One of the extraordinary facets of our modern civilization is that we take lighting fires for granted. All you need is a cheap match. However, this is a relatively recent invention.[1] So how did the early Victorians and their predecessors light their fires? Old films and television series, the so-called costume dramas, rarely, if ever, show people lighting fires. One reason for this was that lighting a fire could be a long process, so once it was lit, it was kept going. Even I remember that letting the fire go out was a heinous crime in the days before central heating, when our house was heated by a coal fire. The fire was kept going during the coldest winter weeks: it was banked up at night with coal, which kept it nicely smoldering while we slept. Its heat prevented the water pipes in the house from freezing during the iciest nights and subsequently bursting when they were warmed up again. It did not heat the whole house though, and I still had to chip the ice off the inside of my bedroom window on winter mornings.

Lighting a fire in a house in the United Kingdom in the 1940s was a substantial enterprise, even with matches readily available. We can only imagine how our great-great grandparents managed it before the invention of matches. And we have only a limited idea of what fire-lighting meant to our ancestors in ancient times. In this chapter, I trace the history of fire-lighting and show, perhaps surprisingly to many readers, that the mineral pyrite was a key to this process, although those with a classical education will already suspect that pyrite had something to do with it, since its name is derived from the Greek and means *firestone*. From fire-lighting, we see how pyrite became the foundation stone for many of the major industries that dominate our world today.

The importance of fire to humankind was recognized by the numerous myths to explain its origin that arose in almost every major culture. Fire was first given to the Greeks by Prometheus, to the Indians by Mātariśvan, and to the Chinese by Sui Ren (Figure 2.1). Hermes actually invented fire, according to Greek mythology, but the Titan Prometheus stole it from Zeus and gave it to humans. He was famously punished for this by being chained to a rock and having his continuously regenerating liver pecked out by an eagle. Mātariśvan found fire, which had been hidden by the gods. His relationship with Agni, the Vedic fire god, is unclear, and he is sometimes described as merely another aspect of Agni. *Agni* means fire in Sanskrit and is cognate with the Latin *igni*. Sui Ren was one of three August Ones, the early mythical emperors of southern China. He "invented" fire, which means that he found a way of kindling it. It seems that each culture had a duality in their fire myths: one entity—usually clearly a god—invented fire in the first place and another one, with at least partial human characteristics, found it or gave it to humankind. This neatly reflects the duality of fire as both a natural phenomenon, as in volcanoes and wildfires, and something humans could create. The partially human entity in the fire myths appears to have been some person from unrecorded time with whom the kindling of fire was later associated.

FIGURE 2.1. The fire gods: the double-headed Mātariśvan from an 18th-century watercolor and the three-eyed Sui Ren (H.C. Bose, *The Dragon, Image, and Demon.* London, 1886).

The ancients recognized that the use of fire was a peculiarly human characteristic that differentiated humankind from other animals. Although fire occurred naturally through, for example, wildfires caused by lightning strikes, the first continual use of fire required a means to generate it. Since early hominids tended to be nomadic, they also needed to find a way of carrying fire-making technology with them.

There were two alternative methods of creating fire in the ancient world. One was by rubbing two pieces of wood together and the other was by striking two stones together. Variations of the wood-on-wood configurations include the fire-saw, whereby one piece of wood is rubbed against a notch in a plank in a sawing motion, and the fire-plough, where it is rubbed against a groove. The friction caused produces enough heat to kindle a fire. The most common wood-on-wood technique method appears to have been fire-drilling using a fire bow. Fire-drilling often involves two people: one rotates the stick while the other presses down on it or catches the fire.[2] It can be made a little easier by looping a bowstring around the stick so that the stick rotates as the bow is moved. In this arrangement, a single driller can produce a wood ember that is then used to light the fire. It has become

popular in modern bushcraft training mainly due to the activities of television survival experts. However, these appear mainly to concern survival
as a form of entertainment. This is different from the real world of pro-
fessionals, like prospectors, who work in extreme environments and need
these survival skills only if something goes wrong. These people tend to
use the easiest and least labor-intensive methods. For example, when the
matches are found to be wet at the end of a tiring day in the field, it is easier
and quicker to use a steel knife on a stone to produce spark than build a fire
bow from wet wood and nettle string. On the other hand, sticks are widely
and freely available, and the fire bow may have advantages in more arid
environments. The television survival expert demonstrates the fire bow in
the Australian outback, not the Swedish Arctic.

Sui Ren, who may have been a tribal chieftain who moved to China
50,000 years ago, is supposed to have invented the fire bow. He is known
as the Fire Driller. Fire-drilling technology appears to have been worldwide in ancient times; fire drill relicts have even been found from the
second millennium BCE in Egypt. Indeed, a decorated fire drill was found
in Tutankahmun's tomb. Figure 2.2 shows a painting of firesticks from

FIGURE 2.2. Who's been a naughty boy then? Hermes with firesticks and his mother
Maia. From an Attic amphora c. 500 BCE in the Staatliche Antikensammlungen of
the Kunstareal of Munich. Photo Bibi Saint-Pol.

6th-century BCE Greece. Fire drills and drilling holes in objects are, of course, closely related, and it may be that the one gave rise to the other. Drilling holes for practical or decorative purposes, such as bead making, has been around for a long time. The bow drills for drilling led to the development of frames and flint-tipped drills for easier and more accurate operations. No doubt the developments in drilling technology also had a feedback effect on fire-lighting methods.

The other method of creating fire was the stone-on-stone method of striking a flint against pyrite.[3] Pyrite was described in the earliest Greek texts, and its name reflects its widespread ancient use in fire-making. *Pyrite* derives from the Greek πυρ, meaning fire, and is a reference to its spark-making properties when struck against flint. In this case, *flint* does not refer only to the hard cryptocrystalline variety of quartz found as nodules in many calcareous sedimentary rocks such as the English chalk but rather encompasses any rock containing quartz that is hard enough to abrade particles from pyrite. This would include many common rock types. On the classical 10-point Mohs scale of hardness, pyrite has a hardness of 6, so that material with hardness greater than 6 is needed to scratch it. Quartz has a hardness of at least 7, depending on its exact form, but is easily hard enough to abrade pyrite.

It is an urban myth that striking two flints together generates sparks suitable for fire-lighting. In fact, the silica particles glow briefly but do not burn long enough to kindle a fire. By contrast, pyrite sparks burn brightly and last long enough to start a fire. In Europe and North America especially, a fire-lighting kit called a *strike-a-light* (Figure 2.3) consisted of pyrite and flint and was often completed with a small bunch of dried *Fomes fomentarius*, a fungus growing inside old or diseased trees. The glowing pyrite particles were caught in this material, which then smoldered. Any other fungus or dry material that will catch a spark will do as well. With careful blowing and the addition of any easily ignitable material, a fire can soon be started. I have done this routinely (with matches instead of pyrite and flint) during fieldwork in the Arctic and can vouch that it works well.

Although the wood-on-wood technique was widely used in historical times by Stone Age peoples, it is probably the more recent method. Ancient Stone Age peoples appear to have mainly used the strike-a-lights. The evidence for this comes from the widespread occurrence of strike-a-light remains in ancient archeological sites and the limited evidence for the wood-on-wood technique. Even though the remains of wooden implements have been found in Stone Age sites, relics of the characteristic

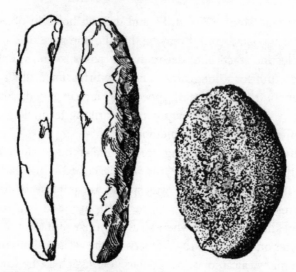

FIGURE 2.3. Bronze Age strike-a-light from the 4,400-year-old cemetery at Seven Barrows, Berkshire, England (Walter Hough, *Fire-Making Apparatus*. Washington, DC: US National Museum, 1890).

wooden tools for fire-lighting, such as those found in Tutankahmun's tomb, are rare. The *Rig Veda* is one of the oldest extant texts in any Indo-European language and dates from 1900–1100 BCE. It contains multiple references to the production of fire both by the wood-on-wood and the stone-on-stone methods.[4] Likewise, modern Inuits of northwestern Greenland and Canada knew of both techniques, although archeological evidence suggests that their ancestors only used flint and pyrite (see endnote 3).

There has been some discussion about whether pyrite or marcasite was the actual iron sulfide mineral used, but this is probably misplaced. The ideal pyrite form for a strike-a-light is a broken rounded nodule of pyrite about the size of a big pebble or small cobble, which can be easily held in the hand, like the one shown in Figure 3.3 in Chapter 3. These nodules of pyrite are often made up of radiating crystals and are commonly called *marcasite* by both amateur mineralogists and professional geologists alike. This is a misidentification, probably relating to the false idea that the elongated and often needle-like shapes of the individual pyrite crystals in the nodule reflect a lower symmetry class in the crystalline material. This led to the idea that the crystals are marcasite, which looks like pyrite and has the same composition but has a lower, non-cubic symmetry. As we see in Chapter 4, this whole idea is mistaken and based on an incomplete understanding of the fundamental nature of crystals. One reason that pyrite has been a key mineral in the evolution of the science of crystallography, as discussed in Chapter 4, is partly because it can occur in a multitude of forms, including needles and even hairs.

Massive lumps of pyrite or individual pyrite cubes do not work as well, probably because they have relatively smooth surfaces and small particles are not easily scraped from them. The radiating crystals of the pyrite nodules produce a rugged surface that chips easily, producing the small particles that fuel the sparks. Pyrite nodules are widely distributed, which makes them even more appropriate for use in fire-lighting. They are commonly found in sedimentary rocks such as limestone, claystone, and shale and are thus easily extractable. Anyone who lives in areas with chalk as the bedrock will be familiar with these pyrite nodules. They stand out against the white chalk, and their presence is often marked by red stains of iron oxide from where the pyrite has oxidized. Analyses of hundreds of these pyrite nodules in the chalk were made in the 1930s and all were pyrite; marcasite was not found. Even with their abundance, it is likely that suitable pyrites for strike-a-lights were treasured possessions in the past and possibly the subject of early trade and barter. Compared with the alternative method of wood drilling, the strike-a-light represented a more sophisticated and easier method of starting a fire. One might suppose that, although wooden sticks were commonplace and freely available, the more sophisticated technology of the strike-a-light would have created a demand that would automatically have led to trade. I can well imagine the fashionable young Stone Age hunter displaying his strike-a-light to his gang in the same way a youth of today might show off the latest cell phone.

The strike-a-light sets themselves became progressively more sophisticated. Thus the flint strikers were worked into round-based forms that could be more easily held. Then flints were fixed into wooden sticks or deer antlers for ease of handling. This then meant that the flint strikers could be formed into shards like knives and ultimately served a dual purpose. A leather pouch was developed to contain the strike-a-light set. These sets had drawstrings so that they could be closed and attached to a belt. This also enabled small amounts of the kindling material such as *Fomes fomentarius* to be carried around and kept dry. In its ultimate form as developed by 18th-century Inuits, a small leather container with some kindling material was included in the set. This had a flap that could be opened to start the kindling smoldering; the flap could be closed to keep it burning in a reduced oxygen atmosphere. The result was a far hotter charcoal fire-starter.

The important thing about the strike-a-lights is that they were portable and could be taken on journeys by traveling peoples. They were thus never without the means of making fires. Dramatic evidence for this comes from Ötzi, the 5,300-year-old mummified man found preserved in ice high in the Alps. He carried a strike-a-light set with him.[5] Strike-a-lights continued in use into the 18th century by the Inuit peoples of northern North America and Greenland, and strike-a-light sets are displayed in several museums in the region.

As far as I am aware, evidence for strike-a-light sets in ancient Egypt and Mesopotamia has not been directly reported. The main problem is the rapid oxidation of pyrite in these climates by contrast with northern Europe, for example, where pyrite may be better preserved in colder regions. However, the use of flint is well evidenced, and flint mines have been widely reported from ancient Egypt. The abundance of flint knives in the remains of these cultures suggests the possibility of similar traveling strike-a-light sets as discovered in more northerly cultures. As a modern woodsman might make a fire with a steel knife and a quartz-rich rock, so the ancient Egyptian or Mesopotamian traveler is likely to have done the same with his flint knife and a lump of pyrite. I must admit that I have not tried the flint knife and pyrite routine but can say from experience that the steel knife and quartz combination works.

After the more widespread availability of steel during the Iron Age, pyrite was replaced by cheaper steel, and steel-based strike-a-lights, often in the form of tinderboxes, were used until the invention of matches in recent times. In this combination, particles of iron from the steel caused the sparks. Every 19th-century home would have had a tinderbox, and one of the first jobs in the morning would be to light the fire for both heating and cooking. I have never tried to light a fire with a tinderbox but I can imagine that it was not a simple task and required skill and experience, as well as time, even indoors. In the wild, the use of steel strike-a-lights and later tinderboxes would have been even more difficult.

The flintlock pistol was based on this process. The flintlock was a later and cheaper development of the wheellock pistol, which was invented around 1500 CE. In the wheellock a steel wheel was rotated against a piece of pyrite and the sparks ignited gunpowder in the pan. This pistol replaced the cumbersome matchlock arrangement where the ignition was enabled by a slow-burning match—basically a fuse. The pyrite-based wheellock is often considered to have been more efficient than the later flintlock pistols since it reliably ignited the powder and fired the shot. The flintlock pistol was especially prone to misfires, as the steel–flint combination did not produce the fat sparks needed to ignite the powder. The modern equivalent of the strike-a-light is the cigarette or gas lighter where the "flint" is actually ferrocerium, an alloy of rare earth metals, magnesium, and iron.

Pyrite was central to humankind's ability to utilize fire. The strike-a-light provided a portable fire-lighting technology to Stone Age people. Being able to light a fire wherever they were enabled nomadic peoples to travel great distances, following the herds of wild animals they subsisted on. Ultimately, it led them out of Africa and into less amenable climates. It enabled them to survive colder and darker northern latitudes where, although game might have been plentiful, the environment was hostile.

The Origins of Human Culture

Charles Darwin stated that the invention of language and the means of producing fire were the two greatest human achievements.[6] However, the cultivation of fire predates *Homo sapiens*. The earliest repeated use of domestic fires, indicating a method of lighting fires to order and not relying on natural events such as lightning strikes, is from a 1-million-year-old cave in South Africa that was occupied by *Homo erectus*. Habitual use of fire is evidenced by the presence of hearths, or deliberately constructed fireplaces. The oldest date from 400,000 years ago and are associated with early *Homo sapiens* and the Neanderthals in Europe, Asia (Israel), and South Africa.[7] The habitual use of fire became common in the middle Paleolithic only 250,000 years ago.

Middle Paleolithic peoples were mostly nomadic and therefore needed to transport fire-making tools. Cooking began some 1.9 million years ago. It has been suggested that *Homo erectus* was adapted to a diet of cooked food and was capable of controlling fire, if not creating it. Eating cooked foods makes digestion easier, and Harvard's English primatologist Richard Wrangham suggested in 2009 that this was a central evolutionary driver toward larger human brains.[8] Cooked food is also healthier, since cooking kills the pathogenic bacteria. Indeed, the Fire Driller, Sui Ren, is venerated today mainly because of his contribution to human health rather than for the invention of the fire bow.

The invention of strike-a-lights meant that fire was available on demand. The consequences were far-reaching for human development, and this was already recognized in the *Rig Veda*. Fire meant that food could be cooked after dark, leaving the daytime free for hunting and other activities rather than eating on the hoof. Fire also provided a light source and prolonged the hours in which humankind could be active. The social consequences were potentially enormous. Groups around campfires could have developed a more abstract form of communication as stories were told and songs sung.

Although the origins of storytelling and singing are at present only accessible by indirect evidence and inductive reasoning, the beginnings of figurative art have a more substantive timeline. Cave paintings in France and Spain and rock art in Australia can be dated more or less accurately. It appears that the beginnings of cave art cluster around 40,000 years ago. One of the earliest key pigments of these ancient artists was charcoal black, the remnants of fires. The development of the means of kindling fires with pyrite strike-a-lights thus helped develop human artistic expression. Interestingly, the other common pigment used by these ancient artists was red and derived from ocher and limonites. These red hematitic ochers are commonly developed from the natural oxidation of pyrite deposits.

A good case can thus be made that pyrite constitutes one of the pillars on which our culture is founded. Pyrite was at the heart of the original development of the arts and sciences and technology. The home fire or hearth was a key feature of daily life since at least 400,000 years ago. The hearth not only attracted many technical tasks, such as cooking, for which fire or heat was needed, but also played an important role in social life. The evolution of a simple language into a complex one, involving abstract concepts, may partly have been driven by the stories told at daily gatherings around the fire. Group and ritual activities associated with the hearth may also have stimulated the development of the arts. It is interesting to consider that the first mineral sought by the ancient prospectors may not have been the exotic gold and silver of later civilizations but everyday pyrite.

The Earliest Chemical Industry

It may not be generally appreciated how important pyrite has been and still is to the world economy and to providing the basics for our current civilization. Pyrite continues to be mined worldwide and is a major source of sulfur, the basic constituent of sulfuric acid. Sulfuric acid has become one of the most important industrial chemicals, and more of it is made each year than any other manufactured chemical.

Sulfuric acid is such an important commodity chemical that a nation's industrial strength can be indicated by its sulfuric acid production.[9] World production in 2004 was about 180 million tons. Sulfuric acid is used in the chemical industry for production of detergents, synthetic resins, dye-stuffs, pharmaceuticals, petroleum catalysts, insecticides, and antifreeze, as well as in various processes such as oil-well acidicizing, aluminum reduction, paper sizing, and water treatment. It is used in the manufacture of pigments and includes paints, enamels, printing inks, coated fabrics, and paper. The list is endless and includes the production of explosives, cellophane, acetate and viscose textiles, lubricants, nonferrous metals, and batteries.[10]

Sulfuric acid is a relatively recent manufactured chemical. The oldest reference to the substance is generally accepted to be in the writings of the anonymous medieval alchemist known as Pseudo-Geber. Prior to this, the important analogous chemical substances were the sulfate salts of iron, copper, and aluminum, known to the ancients as the *vitriols*. These occurred in the lists of minerals compiled by the Sumerians 4,000 years ago, which are discussed in more detail in Chapter 3. They are important because they were used as mordants in the dyeing industry. In order for natural dyes to be fixed in the cloth—and not be washed out during the next rainy day—it is necessary to treat the cloth with a mordant. The mordants widely used in dyeing were solutions of the vitriols.

Vitriols occur naturally, if rarely, as precipitates from volcanic fumaroles, as a component of salts in gypsum deposits, and as oxidation products of pyrite. For example, Galen, the great physician of 2nd-century CE Rome, described vitriol being collected in Cyprus, an island famous in antiquity and even today for its abundant copper-rich pyritiferous ore deposits. Indeed, the word *copper* derives from the Latin *aes Cyprium* (metal of Cyprus), which was later shortened to *cuprum*. Galen reported that green iron sulfate solutions produced from weathering of the ores were drained into a warm cave where the crystals of iron sulfate were crystallized. Some classicists have wrongly identified these crystals as copper sulfate, but these would have been blue.[11] The most important of the ancient vitriols was alum, a complex double salt of potassium aluminum sulfate. Alum occurs naturally in alunite, a rock associated with volcanic fumaroles, and the alums are constituents of some gypsum deposits. Alums were produced at sites of volcanic alunite at Melos and Lesbos in

Greece from the 2nd century CE to the 7th century CE. Gypsum deposits in the eastern deserts of Egypt were important sources of alums from 1500 BCE, and the use of alum as a mordant for dyeing cloth red with madder can be traced back to this time. However, these natural alums are not pure. In fact, they are not pure enough for their indiscriminate use either as mordants or, as discussed later, for medical purposes.

The demand for quality mordants in cloth manufacture rapidly outstripped any potential natural source. The problem was that differently colored cloths were not only aesthetically desirable but became signals of social status as, for example, in the purple togas of the Roman emperors and the yellow robes of their Chinese counterparts. Thus large quantities of variously colored cloths were required. The demand for vitriols could not be satisfied from natural supplies, and industries developed to manufacture this substance from pyrite. The processes involved in their manufacture involved a series of steps, including the mining of pyrite, the controlled oxidation of pyrite, and the purification of the product. This latter was a key to the preference for the manufactured product over the natural material: the natural alums were impure, and the dyeing industry required a pure material. The steps in the manufacture of mordants were carried out in sequence by a series of different craftsmen and transported between sites in an unfinished form before being finally completed and sent on to the sales site.

In 1637, Song Yingxing reported how sulfur and green vitriol were prepared from pyrite in China.[12] About 250 kg of pyritiferous shales were surrounded by 500 kg of coal and a hole was left in the top of the pile. The pile was ignited at the base and left to burn for ten days. The yellow vapor that came out of the top of the pile could be condensed to produce sulfur by the simple expedient of covering the top of the pile with an earthenware bowl (Figure 2.4). The cinders were removed and dumped in water for three hours, then filtered and boiled down to one-tenth of its volume. Green vitriol crystallized out at the top. This process was itself the basis of a series of major industrial developments.

The importance of the vitriols in ancient China can be gauged by the fact that there was no single Chinese word for *pyrite*. Rather the mineral names were related to the type of vitriol the pyrite produced. Thus pyrite extracted from coal deposits was called *meikuang shao fanshi* (煤矿烧矾石) or *coal vitriol*. The Chinese word for vitriol is *fan*, which may also be translated as *alum* or *alum-like substance*. Several types of vitriol were manufactured from the basic green vitriol or hydrated iron sulfate, including yellow

FIGURE 2.4. Production of sulfur from pyrite in China. The pyrite was placed in a brick kiln, heated for ten days, and the "golden vapors" collected as sulfur in an upturned bowl on the top. (Apparently from an original illustration by Song Yingxing, *Tian Gong Kai Wu*, 1637. I could not find this image in the original Chinese edition of the *Tianging Kaiwu* from the Chinese National Library. The citation comes from Y. Zhang. 1986. Ancient Chinese manufacturing processes. *Isis*, 77:487–497 and the image is reproduced by permission of Chicago University Press).

vitriol, which was another, more oxidized form of hydrated iron sulfate; black vitriol, which was a mixture of green vitriol and iron sulfide; and red vitriol, which was a suspension of iron oxides in water.[13]

The most valuable product was alum, or white vitriol, which is a double salt of aluminum sulfate, usually with potassium or ammonium.[14] Potash was obtained from burning seaweed and ammonium from urine. Boiling the liquid with potash or urine was used to crystallize the alum and separate this from iron sulfate. This was the key to the process, and the skill of the alum maker was in the judgment of the exact point when the alum crystallized but the iron sulfate remained in solution and could be poured away. Contamination with iron sulfate would cause the dyed cloth to blacken, so only pure alum could be used as a mordant for colored cloth. This process was used until self-fixing dyes and synthetic sulfuric acid manufacture were discovered in the 1800s.

It is unknown who invented the production of alum from organic-rich pyritiferous shales or alum shales or where it originated. It became particularly important in late medieval England when much of the nation's wealth was dependent on the wool trade. The dyeing of wool required that the mordant used be white, and this in turn required good-quality alum. The alums had been imported to England from natural sources in the Middle East and, in the 15th century, from Italy. During Tudor times the supply from the Papal States became less certain, especially after the excommunication of Henry VIII. As the wool trade continued to expand, the demand for alum made the development of a home-based production method imperative. However, the problem was not successfully resolved until 1604 when Thomas Atherton and his wife Katherine opened the first alum works on their estate in Yorkshire in northern England.

The process developed for the manufacture of alum from alum shales in Europe was similar to that described for sulfur and vitriol production in China. The difference was that the presence of organic matter in the alum shale improved its flammability so that burning could be initiated by brushwood. There was no need for coal to provide long-term burning since the organic matter in the shale provided the fuel. A clamp of alum shale mixed with brushwood some 10 m high and 50 m long was burned for months. The burnt shale was steeped in water to dissolve the sulfate produced from the oxidation of the pyrite. The clay minerals in the shale provided the aluminum and pyrite oxidized to sulfate, which, when dissolved in water, produced sulfuric acid. Alum was produced by the reaction of the aluminum components of the clays and the sulfate from the sulfuric acid.

The result of the demand for alum was the foundation of an industry in Yorkshire, England, to process alum shale and produce alum. The alum works on the North Yorkshire coast produced alum from alum shale for almost 250 years between 1604 and 1852. Huge quantities of pyritiferous alum shale outcropped here and were mined from large quarries. The coastal location also allowed urine, seaweed, and coal to be brought in by boat. The burnt clamps of shale were steeped in water in large pits to produce the aluminum sulfate. The coal was used to boil the aluminum sulfate liquor. Burnt seaweed was added to produce potash alum or urine to make ammonium alum. The key process was the sequence of repeated washing and recrystallization of the alum, and this was kept so secret that no contemporary accounts of the method remain. This process produced the high-grade, exceptionally pure alum that was much sought after by the dyers.

Pyrite in organic shale in Yorkshire played a basic role in initiating the Industrial Revolution. The alum industry was based in a remote area of England, and the combination of large industrial sites, including quarries and factories, with the requirement for long-distance transport helped drive the development of industry as we know it today. In 1845 Peter Spence, a Scottish entrepreneur, discovered that he could manufacture alum more cheaply and on a larger scale by treating cheap shale waste from coal mines with commercially produced sulfuric acid. Even though this, together with the development of aniline dyes, heralded the end of the alum shale industry, the sulfuric acid used was still produced from pyrite.

The production of pure alum from pyrite has been described as the point of origin of the modern chemical industry.[15] The reason is that the process required not only the manufacture of a chemical substance but also its purification. Alum was the first substance to deliberately be prepared in a substantially pure state as we would understand it today.

Big Pharma

The manufacture of artificial drugs—in contrast to the use of natural remedies—can be traced back to pyrite and strike-a-lights. It is not a big step to drop pyrite from a strike-a-light into the fire. The result is the formation of sulfur oxide gases with their characteristic burnt smell. These sulfur oxide gases, apart from being poisonous in high doses, can clear clogged-up noses and are very useful in fumigation. In other words, the ancients had a direct pathway into potential medicinal remedies, as well as a method of cleaning up bug-infested cloths and animal skins. Elemental sulfur itself condenses as a bright yellow sublimate on the cooler areas of the fire, although this result is best achieved in closed hearths with limited access to air. Sulfur was used for fumigation in preclassical Greece.[16] Homer described Odysseus using sulfur to fumigate the palace after he had killed Penelope's suitors, and its "pest-averting" properties are mentioned in the *Odyssey*.

Deposits of native sulfur are widespread in southern Europe, where they are associated with volcanic fumaroles. In Egypt and the Middle East, native sulfur is found in gypsum. Sulfur was certainly known to the Chinese by 600 BCE,[17] even though China has only a few minor reserves of natural sulfur and these are mainly located in border and remote areas. The challenge of transporting natural sulfur from these remote areas restricted its use, and there is much evidence that, even in ancient times, the main source of sulfur in China was pyrite. By 300 CE sulfur was being produced from pyrite in Shanxi, Hebei, Henan, Hunan, and Sichuan provinces. One of the earliest descriptions of the medicinal use of sulfur was in the *Shen Nong Bencaojing* or *The Pharmacopeia of the Heavenly Husbandsman* compiled in the Western Han period (206 BCE–24 CE), which catalogued the medicines invented some 3,500 years earlier by the legendary emperor Shen Nong.[18]

The importance of pyrite to the pharmaceutical industry was mainly related to the production of sulfur for medicinal purposes. As with the manufacture of alum, discussed earlier, the medical sulfur had to be produced from pyrite in the absence of deposits of natural sulfur. Sulfur was used, mainly in creams, to alleviate conditions such as scabies, ringworm, psoriasis, eczema, and acne. The mechanism of action is unknown—though sulfur does oxidize slowly to sulfurous acid, which in turn (through the action of sulfite) acts as a mild reducing and antibacterial agent. According to the *Ebers Papyrus*, an important medical papyrus from 1550 BCE, a sulfur ointment was used in ancient Egypt to treat granulated eyelids (*blepharitis*). The *Ebers Papyrus* was based on now-lost earlier works possibly dating from 3400 BCE. A similar prescription was inscribed on a clay tablet from 600 BCE Nineveh. Wei Bo Yang, in the *Can Tong Qi*, possibly the earliest book on alchemy dating from 142 CE, recorded several recipes that included both spiritual (the cultivation of "vital energy") and chemical medicines (particularly combinations of lead, mercury, and sulfur) to maintain health and prolong life. Its full title is the *Zhou Yi Can Tong Qi* or *Token for the Agreement of the Three in Accordance with the Book of Changes*. This identifies its ancestry through the *I Ching* or *Book of Changes* and thus directly to the older *Classic of Rites* mentioned earlier. These recipes seem to have worked, since Wei Bo Yang lived past age 70, even though a key constituent of his medicines was mercury, an element that is regarded as highly poisonous today.[19]

In the ancient Mediterranean and Near Eastern civilizations, sulfur was found naturally and did not need to be manufactured from pyrite. Even so, pyrite was used as a basis for the manufacture of some medicines. The situation is somewhat confused since, as mentioned in Chapter 3, the definition of the mineral in ancient medical texts is shrouded in imprecision and seems to have encompassed both zinc and copper sulfide smelter slag.[20] We are on firmer ground with the use of pyrite as a basis for products such as the vitriols and alums manufactured for the dyeing industry as described previously. Iron sulfate or green vitriol was a useful additive in certain medicines in mixtures to treat pustules on infants and as an emolument for scrofulous tumors. The use of alum in medicine has been documented for more than 2,000 years since the Babylonians listed it in one of the first pharmacopeias. The main medicinal use of alum was, as it still is today, as an astringent to improve wound healing. The modern styptic used to close up razor nicks occurring after wet shaving is alum-based. It has also been used as an emetic to treat someone who has ingested a poison. The direct application of powdered alum to open wounds and sores was also widely recommended by the ancients; however, direct application to open wounds can be painful. Its use in helping reducing swelling of the skin around healing sores appears more appropriate. The alum used for medical purposes had to be even purer than that needed in the dyeing industry. Its manufacture must have required the highest degree of skill by the alum makers, particularly in the level of temperature control required—in the absence of thermometers—for the repeated dissolution and crystallization that was central to the purification process.

Sulfur was a key ingredient in early medicines, and it was manufactured in part from pyrite. The manufacture of sulfur-containing drugs and fumigants is at the root of the development of nonvegetable pharmacy. The ancients—like us today—had a pharmacopeia largely based on organic compounds with specific medicinal properties. However, the use of inorganic sulfur-based compounds, which was especially developed in Asia, required a deliberate manufacturing line. Pyrite needed to be mined, burned in closed hearths, and the resultant sulfur mixed into the required unguents and pills. The manufacture of purified products such as the vitriols and alums from pyrite required the development of a series of skilled operations.

Pyrite and the Origins of the Modern Arms Industry

Daoism is aimed at finding a way to immortality. This may be achieved spiritually but also with the aid of elixirs, the ultimate being the elixir of life. Chinese Daoist alchemists and monks around the beginning of the Current Era began to record their ideas about how the way (*dao*) might be found. Shen Nong, the Heavenly Husbandman, is reported to have described *xiaoshi* (saltpeter or potassium nitrate) as the highest of the three grades of *materia medica* in his classical pharmacopeia of the 2nd century BCE.[21] It is therefore unsurprising that the ancient Chinese alchemists tried mixing saltpeter with a variety of other components, such as the sulfur derived from pyrite, to produce the elixir of life they sought. They soon found out that mixing sulfur, cinnabar, saltpeter, and organic materials was quite dangerous and produced spontaneous ignition if not minor explosions. Indeed, there are distinct warnings in Shen Nong's pharmacopeia to avoid these mixtures, which "fly and dance" in a violent reaction.

It could be argued that Shen Nong was describing fireworks in his pharmacopeia, and there is a good case that the early Chinese alchemists knew of a splendid parlor trick to produce flames and explosions. By around 300 CE, the alchemist Ge Hong had recorded the components of gunpowder: saltpeter, sulfur, pine resin, and certain carbonaceous materials in the *Bao Pu Zi* or *Book of the Master of the Preservations of Solidarity*.[22] Firecrackers, packages of explosives strung together with string, were known in the 4th century. Sun Si Mao, one of the fathers of Chinese medicine, refined the recipe for gunpowder around 600 CE. The modern Chinese credit Li Tian, a 7th-century CE alchemist, as the inventor of fireworks.[23] The story goes that Li Tian ignited a bamboo tube filled with an explosive mixture in order to rid the emperor Li Shiming of a dragon that was annoying him. He also set off fireworks to help people in eastern Hunan with floods and droughts.

The containment of the explosive nature of gunpowder remained a problem for the ancient Chinese. There are many accounts of injuries and uncontrolled explosions in the early literature.[24] Early Chinese gunpowder was low in saltpeter content and thus burned fiercely rather than exploded. However, Chinese military engineers realized the obvious military potential of black powder and by 904 CE were hurling lumps of burning gunpowder with catapults during a siege. Flame-throwers, black powder-based weapons fueled by a stream of naptha, were deployed in 919 CE. There is a wonderful painted silk tapestry from 10th-century China in the Musée Guimet, Paris, showing the Buddha being attacked by Mara the Temptress and her demonic hordes. The Buddha, who is some ten times larger than the demons, is depicted sitting serenely in the center unaffected by the surrounding mayhem. In the right-hand quarter of the silk banner, one large demon in a loincloth and with a postmodern blue tattoo on his stomach hurls a flaming gunpowder bomb; nearby a large white demon with three snakes growing out of his head is operating a gunpowder-fueled fire lance. The banner was discovered in 1900 together with a hoard of other manuscripts in the Mogao Caves of Dunhuang, a city in western China on the ancient Silk Road. The military potential of gunpowder had reached Dunhuang by the 10th century, and the artist, probably a Buddhist monk, was fully aware of the latest military technology. By 969 CE gunpowder-strapped arrows were being produced and by 1044 CE gunpowder bombs. The fire lance was a proto-gun, and a copper gun dated from 1271 was discovered in Ningxia province in 2004. By the end of the 13th century, hooped iron cannons were being produced.

This gunpowder-based weaponry spread westward, probably along the Silk Road, and the Arabs began producing rockets and fireworks.[25] By 1382 CE, Arab armies were using gunpowder-based weapons in sieges. Roger Bacon first mentioned the recipe for gunpowder in 1267.[26] In 1346 CE, King Edward III employed gunpowder weapons against the French at the battle of Crecy. The internecine wars following the demise of the Mongol Empire in China led to a further arms race that culminated in the deployment of ship-borne cannon at the Battle of Lake Poyang in 1363 CE.

The basis of the modern arms industry had been established worldwide by the 14th century. Although saltpeter was a key part of this process, pyrite-derived sulfur was integral since it lowers the ignition temperature, burns at a temperature above the ignition temperature for saltpeter, and increases the combustion speed. Gunpowder manufacture began to dominate the industrial use of sulfur so that, for example, 200 to 400 tons of sulfur were being produced each year for gunpowder manufacture in 18th-century China, mainly from pyrite nodules in coal (see endnote 17).

Throughout Asia, the Middle East, and Europe, mass production of gunpowder and gunpowder-based weapons was in progress in the late Middle Ages. This ultimately led to the Industrial Revolution at the end of the 18th century as the British were locked in a life-or-death struggle with Napoleonic France and needed more cannon and more iron. The industrial-military complex was established. In all this pyrite played a key role and it has continued to be a primary material for the arms industry through the middle of the 20th century.

Pyrite Feeds the World

We have seen that pyrite is the raw material from which sulfuric acid can be made, and a major use of sulfuric acid in modern economies is in the production of fertilizers. About 60% is currently consumed for fertilizer manufacture, especially superphosphates, ammonium phosphate, and ammonium sulfates.

On July 1, 1843, John Bennet Lawes, an English agricultural chemist, first sold his new and improved patented superphosphate fertilizer. In contrast to earlier attempts, this product rapidly released the key nutrient phosphorus into the soil in a form that plants could readily metabolize. Lawes' process was based on the reaction between sulfuric acid and phosphate rock. The result of Lawes' innovation was the increased demand for sulfuric acid. By 1871 there were more than eighty factories in the United Kingdom producing superphosphate fertilizer and shipping it around the world.

During the early part of the Industrial Revolution, sulfur in Europe was sourced from natural sulfur deposits associated with volcanic fumaroles in Sicily. In 1839 the Sicilian deposits came into the hands of a French company, which raised the price threefold. This led to sulfur users in other countries reverting to pyrite as a source of sulfur. Roasting of pyrite produces sulfur oxide gases, and these can be dissolved in water to produce sulfuric acid. Byproducts of the process include copper metal from the pyrite and an iron-based slag that is used in road-building.

The scale of the importance of pyrite to global food production can be appreciated by examining the increase in production of sulfuric acid during postindustrial times. The key point here is that the Industrial

Revolution was constrained by the amount of food needed to feed the bur-geoning population. For example, it has been estimated that the population of Britain was constrained to around 6 million in preindustrial times due to the limitations of agricultural productivity.[27] This compares with over 60 million today. The excess 54 million people are fed by postindustrial technological advances. This step increase in agricultural productivity was fueled by the development of industrial fertilizers. This, in turn, caused a consequent exponential increase in the demand for sulfuric acid, sulfur, and pyrite.

Pyrite continued to be the main source of sulfur until Herman Frasch developed a hot-water process for extracting sulfur from the extensive underground deposits of native sulfur associated with salt domes on the US Gulf Coast in 1894. However, by 2004 no sulfur was produced in the United States by this process: sulfur recovered from sour gas and petroleum was the dominant source, driven by both economic and environmental reasons.

The cost of transport and strategic considerations meant that pyrite-derived sulfur was still responsible for about half of the world's sulfur until the middle of the 20th century. Pyrite reserves are distributed throughout the world, and known deposits have been mined in about thirty countries. Currently global pyrite production is about 14 million tons per year, and about 85% of this is produced in China.[28] Pyrite is mainly used to produce sulfuric acid and is equivalent to around 7 million tons of sulfur annually with a value of about US$160 million. This amounts to around 10% of the total world sulfur production. Most of this sulfur is used in sulfuric acid manufacture, and most of the sulfuric acid is used to make fertilizers. In this context, pyrite continues to be a major factor in food production.

Pyrite, Civilization, and Culture

The common thread in the role of pyrite in the early development of our civilization and culture is that it facilitated activities that, although possible without its use, would have been far more inefficient and limited. Its contribution to the taming of fire is often overlooked but must have been central to the everyday life of early humankind. The idea that you could light a fire wherever you are is obviously fundamental to your lifestyle. The cultural consequences were even more significant than the washing machine and dryer replacing the tub and mangle was to recent humans. Instead of needing to stay in the house to keep up with the daily necessities of washing and cooking, these conveniences released a large portion of the human race into more constructive activities while at the same time improving the quality of life and health. Pyrite gave the human race quality time. The consequent cultural developments included more time for entertainment and these are a major part of our current everyday experience.

The use of pyrite increased humankind's efficiency. Another way of thinking about this would be that pyrite is the universal common ancestor of technology. The use of pyrite to produce fire is fundamental to all our basic secondary industries but also had a feedback to our primary industries, such as enabling the smithy to produce ploughshares and rock hammers.

Pyrite was also involved in the earliest manufacturing, where things were deliberately made through a process of consecutive activities, like the production of alum. Although a number of animals are now known to use tools in their everyday activities, the development of manufacturing processes involving a sequential series of related activities, often by different individuals with different skills, appears to be peculiarly human. The resultant specializations, divorced from the everyday need for food gathering, must have made a substantive contribution to the development of economics, as barter evolved into the use of money as an exchange medium.

Pyrite has also contributed fundamentally to enabling the human population to expand to more than 7 billion individuals. The human race is going through what is termed in microbiology a *logarithmic growth phase*. Microorganisms are commonly cultured in the laboratory in batch cultures, for example, in a test tube, where the supply of nutrients necessary for their growth is limited to the amount added to the test tube at the beginning of the experiment by the operator. The organisms go through a lag-phase, where growth is low, and then enter a logarithmic phase, where the growth rate increases by doubling at regular intervals. This plateaus out to a relatively stable phase, where the number of individuals is kept fairly constant by a balance between the population and the availability of nutrients. Finally, the population begins to decline as the nutrients are used up. This decline is asymptotic: it never reaches zero since the dead organisms themselves provide a limited supply of nutrients for a small population.[29]

The analogy with the human population and the Earth is unmistakable, although I hope that cannibalism or necrocannibalism is not going to happen during the declining phase. The Earth is our test tube, with a limited supply of nutrients, and the population has undergone exponential growth since the Industrial Revolution. According to the United Nations,[30] the population is expected to level out at around 10 billion during the latter part of the 21st century and then decline at least until a new source of nutrients is found. One of the prime drivers of the current exponential growth phase has been pyrite. In this sense, pyrite is fundamentally responsible for the technical and social changes we have experienced in the modern world. It is at the core of our civilization.

Notes

1. An early dangerous and expensive self-igniting match was invented by Jean Chancel in 1805, but the friction match was invented in 1826 by John Walker, an English chemist.

2. Photograph by K. Birket-Smith. 1927. *Eskimoeme* (Copenhagen: Gyldensdalske Boghandel), 121pp. A splendid line drawing of this photo by Lykke Johansen is more readily available in D. Stapert and L. Johansen. 1999. Flint and pyrite: Making fire in the stone age. *Geologie en Mijnbouw*, 78:147–168.

3. D. Stapert and L. Johansen. 1999. Flint and pyrite: Making fire in the stone age. *Antiquity*, 73: 765–767. See also W.M.F. Petrie. 1891. *Illahun, Kahun and Gurob, 1889–90* (London: David Nutt).

4. See *The Hymns of the Rigveda*. 1896. Translated with a Popular Commentary by T.H. Ralph Griffith. Edited by J.L. Shastri. Vol. 1, new rev. ed. 1973. (New Delhi: Motilal Banarsidass), 707pp. W. Doniger. 1981. *The Rig Veda: An Anthology: One Hundred and Eight Hymns, Selected, Translated and Annotated* (New York: Penguin Classics) v. 402.

5. M. Egg, R. Goedecker-Ciolek, W. Groenman van Waateringe, and K. Spindler. 1993. Die Gletschermumie vom Ende der Steinzeit aus den Otztaler Alpen. *Jahrbuch des Romisch-Germanischen Zentralmuseums*, 39:3–128.

6. C. Darwin. 1871. *The Descent of Man and Selection in Relation to Sex* (London: John Murray).

7. F. Berna, P. Goldberg, L. Kolska Horwitz, J. Brink, H. Holt, M. Bamford, and M. Chazan. 2012. Microstratigraphic evidence of in situ fire in the Acheulean Strata of Wonderwerk Cave, Northern Cape Province, South Africa. *Proceedings of the National Academy of Sciences USA*, 109:E1215–E1220. W. Roebroeks and P. Villa. 2011. On the earliest evidence for habitual use of fire in Europe. *Proceedings of the National Academy of Sciences USA*, 108:5209–5214. V. Aldelas, P. Goldberg, D. Dandgathe, F. Berne, H.L. Dibble, S.P. McPherron, A. Turq, and Z. Rezek. 2012. Evidence for Neanderthal use of fire at Roc de Marsal (France). *Journal of Archeological Science*, 39:2414–2423.

8. R. Wrangham. 2009. *Catching Fire: How Cooking Made Us Human* (New York: Basic Books).

9. P.J. Chenier. 1987. *Survey of Industrial Chemistry* (New York: Wiley), pp. 45–57. Justus von Liebeg stated in 1843 that "We may fairly judge of the commercial prosperity of a country from the amount of sulphuric [*sic*] acid (oil of vitriol) it consumes" (*Familiar Letters in Chemistry* [London: Taylor and Walton]).

10. W.G. Davenport and M.J. King. 2006. *Sulfuric Acid Manufacture: Analysis, Control and Optimization* (Amsterdam: Elsevier). N.N. Greenwood and A. Earnshaw. 1997. *Chemistry of the Elements*, 2nd ed. (Amsterdam: Butterworth–Heinemann).

11. K.G. Kuhn. 2011. De simplicium mendicamentorum temperamentis ac facultatibus lib IX. *Claudii Galeni Opera Omnia*, 12:159–244.

12. S. Yingxing. 1637. *Tian Gong Kai Wu* (The Exploitation of the Works of Nature). See J. Needham. 1986. *Science and Civilization in China: Vol. 5, Chemistry and Chemical Technology* (Taipei: Caves Books).

13. Ho Pen Yoke. 2007. *Explorations in Daoism: Medicine and Alchemy in Literature* (London: Taylor and Francis), 228pp., notes that twenty-two different types of alum or *fan* are listed in the *Zhengtong Daozhang*, the 15th-century Daoist canon, alone.

14. Hydrated ferrous sulfate is $FeSO_4.7H_2O$; hydrated ferric sulfate is $Fe_2(SO_4)_3.9H_2O$; white vitriol or alum is potassium aluminum sulfate $KAl(SO_4)_2.12H_2O$ or ammonium aluminum sulfate $NH_4Al(SO_4)_2.12H_2O$.

15. C. Singer. 1948. *The Earliest Chemical Industry* (London: Folio Society).

16. G.R. Rapp. 2009. *Archaeomineralogy* (Berlin: Springer), 242pp.

17. Y. Zhang. 1986. Ancient Chinese manufacturing processes. *Isis*, 77:487–497.

18. Shen Nong, the Heavenly Husbandsman, was a semi-mythical emperor who lived around 3,700 years BCE. He introduced agriculture and personally tasted hundreds of plants to test their medicinal values. He also introduced acupuncture.

19. J. Godetsky and J. O'Brien. 2009. Mysticism and urology in ancient Egypt. *Urology*, 73:476–479. G.E. Smith. 1930. *The Papyrus Ebers*. Translated by Cyril P. Bryan (London: Garden City Press), 167pp. See discussion about the development of the *Cantong Qi* text and its implications for the actual date of the origins of Chinese alchemy in F. Pregadio. 2002. The early history of the Zhouyi cantong qi. *Journal of Chinese Religions*, 30:149–176.

20. Aulus Cornelius Celsus. *De Medicina*. Book 5, Chapters 18 and 28. Pedanius Dioscorides. *De Materia Medica: Being an herbal with many other medicinal materials*. Translated by Tess Anne Osbaldeston, 2000. (Johannesburg: Ibidis Press). Aelius Galen. 1453. *Method of Medicine*, Vol. 3, Book 13. Translated by Ian Johnston and G.H.R. Horsley, 2011. (Loeb Classical Library). In Book 5, Sections 84 and 153, Dioscorides refers specifically to *cadmia* and *diphryges*. *Diphryges* has been mistakenly translated as pyrite but actually describes forms of furnace slag derived from smelting copper in Cyprus. Its literal meaning includes the idea of

twice burnt. Cadmia is an old name for calamine and a zinc oxide product of zinc smelting. Dioscorides notes that *cadmia* is sometimes produced from pyrites dug out of a hill that lies over Solis. But he notes that this deposit also includes a number of other materials.

21. Although there are deposits of saltpeter in southwestern China, most of the salt-peter was harvested as a seasonal efflorescence from the decay of nitrogenous organic matter. Saltpeter was harvested from the fields, from the walls of houses, and even from dust swept up from the floor. Tao Hongjing described the medici-nal uses of saltpeter and the way in which it was obtained. For example, saltpeter was used to aid wound healing: it was added to the wound and then washed in (or even licked in by cows) to facilitate nitrite production and to aid in sterilization. The Chinese alchemists knew how to identify this salt by a simple flame test: it burns to produce purple smoke; this was described by Tao Hongjing (456–536) in the *Mingyi Bella* (Informal Records of Eminent Physicians). It is thought that saltpeter was used in ancient Chinese medicine because it is a source of nitrite, either through burning or by reactions with the tongue on ingestion. Nitrite is now known to be a valuable treatment for a variety of conditions, not the least of which is angina. Shen Nong recommended the long-term ingestion of nitrate as a contribution to a longer life and ultimate immortality. Because of its bur-geoning military importance, Chinese authorities tried to restrict the spread of saltpeter: in 1067 an edict forbadd the selling of saltpeter or sulfur to foreigners and in 1076 all private transactions in these materials were banned. Saltpeter was also known as Chinese snow in the West. See Needham. 1976. *Science and Civilization in China: Vol. 5* (Cambridge UK: Cambridge University Press).

22. See J. Needham. 1976. *Science and Civilization in China: Vol. 5. Bao Pu Zi* was actually Ge Hong's Daoist pseudonym and literally means the "Master who Embraces Simplicity."

23. Li Tian's birthday (April 18, 601 CE) is celebrated in China today with fireworks.

24. For example, Chung Yin, *Chen Yuan Miao Tao Yao Lueh* (Classified Essentials of the Mysterious Tao of the True Origins of Things), a Tang Dynasty book written around 850 CE, notes that "Some have heated together sulfur, realgar and salt-peter with honey; smoke and flames result, so that their hands and faces have been burnt, and even the whole house where they were working burnt down."

25. The earliest Arab report of the recipe for gunpowder was Hassan al-Rammah Najm al-Din al-Ahdab. c. 1280 CE. *Kitab al-Furusiya wa'l-Munasab al-Harbiya* (Treatise on Horsemanship and Stratagems of War).

26. Roger Bacon wrote to Pope Clement: "a child's toy of sound and fire made in various parts of the world with powder of saltpetre, sulfur and charcoal of hazelwood."

27. See, for example, M. Overton. 1996. *Agricultural Revolution in England: The Transformation of the Agrarian Economy 1500–1850* (Cambridge, UK: Cambridge University Press), 206pp.

28. See L.E. Apodaca. 2011. Sulfur. In *USGS Minerals Yearbook* (Washington, DC: US Department of the Interior), 15pp.

29. I have happily recultured batch systems after two years, for example. That is, even after two years of apparent stasis the organisms will start to develop logarithmically again if a new supply if nutrients becomes available.

30. United Nations. 2004. *World Population to 2300*. Report ST/ESA/SER.A/236. (Washington, DC: United Nations Department of Economic and Social Affairs, Population Division).

3

What Is Pyrite?

The Importance of Names

In this chapter I show how pyrite was at the heart of our early understanding of the composition of substances and how it was central to the acceptance of the revolutionary idea that substances have fixed compositions. This, in turn, was the evidential basis for the modern atomic theory.

Taxonomists will argue that naming things accurately is important since otherwise no-one will know what you are talking about. They would disagree with Shakespeare that a rose would smell as sweet whatever it was called on the grounds that you would not know that a rose was being described. Even so, the only reason things are named is because of need. Thus Homer did not have a word for *blue* because he never needed it: the blue sea became wine-dark, for example. By contrast, contemporary ancient Egyptians had a word for *blue* because they used the blue mineral lapis lazuli for decoration.

The mineral pyrite has been employed by humankind for millennia, and it needed a name. Its long history means that a variety of terms have been used to describe it, often reflecting the technology available at the time. In order to understand the role that pyrite has played in the past, we need to interpret the various names given to this mineral by earlier authorities. This problem is compounded since its history is determined by ancient texts and these were commonly written down by scribes from direct dictation. The scribes rendered the sounds of words as best they could within the limitations of the current orthography. Before the advent of printing, copyists made reproductions of these original texts according to the customs and mores of their local culture. The texts that have come down to us are usually the result of the work of several generations of copyists, and the interpretations become like a game of Chinese whispers. Whether or not a word in an ancient text means pyrite is, at best, a matter of relating it to a description that reflects key properties of the mineral. At worst it may mean probing the etymology of the word and considering its context.

Even today, pyrite is often described in the general literature as *pyrites* or *iron pyrites*. The use of the plural to describe the mineral is a little strange and derives from its history. *Pyrites* was used to describe a group of metal sulfide ore minerals. Metal sulfide ores are usually massive, often brassy yellow, and individual mineral crystals cannot be distinguished easily with the naked eye (Figure 3.1). These ores are usually dominated by pyrite, but the crystals have no obvious regular shape and are mostly intergrown with each other and with other metal sulfides. The modern mining geologist uses a hand lens to inspect the ores, and, even then, a microscope may be required to see the individual minerals.

The title of this chapter is derived from a loose translation of a chapter in Henckel's *Pyritologia: Was ist Kieß* published in 1725. Henckel used the word *kies* in the subtitle of his work. *Kies* is a German miner's term for sulfide ores that are usually dominated by pyrite. The important thing here is that the iron in pyritic sulfide ores is not the major economic target: these ores are primary sources of copper, lead, zinc, sulfur, arsenic,

FIGURE 3.1. Massive pyrite: part of a wall of golden massive pyrite several meters high from Udden, northern Sweden (see color plate).

silver, and gold. The German term *kies* and its derivatives in other northern European languages are still used by miners for metal sulfide ores that contain these valuable metals, as well as iron.

The involvement of the word *iron* in *iron pyrites* seems superfluous today since it is common knowledge that pyrite is an iron sulfide. It reflects the older usage of the term *pyrites* and harks back to the period around 1700 CE when it was first reported that pyrite contained iron. Thus *iron pyrites* are pyrites with iron as a dominant component; this characteristic distinguishes these pyrites from other metal sulfides, especially *copper pyrites*, for example, which may be used for sulfide ores with abundant chalcopyrite, a copper iron sulfide that is the main ore for copper. *Magnetkies*, an alternative name for pyrrhotite, contains some varieties that are magnetic, and *wasserkies* (a corruption of *weisserkies*, white ore) is a German miner's term for marcasite. *Pyrites*, even in the fairly recent past, referred to a group of sulfide minerals and not necessarily to pyrite as we define it today. In many ways, this is unsurprising, since pyrite is defined today in terms of its chemical composition and crystal structure, both of which were unobtainable until the technological advances of the past 200 years. As we look deeper into historical time, the identification of pyrite becomes murkier, but the mineral is outstanding in its color, luster, and common occurrence as large, quite gaudy crystals, as well as its usefulness for fire-lighting and as a source of sulfur. With a little care we can trace back its history with some certainty.

Classical Origins

Pyrite, as a spark-forming mineral, was described in the earliest science texts. Its modern name derives from the Greek πυρ, meaning *fire*. Pyrite occurs in the oldest literature: the 4,000-year-old cuneiform lists on clay tablets produced by the Sumerians. The Sumerians were very much involved in compiling word lists, since they derived omens from observations of natural phenomena. The Sumerian for stone was *za*, and it was common practice for Sumerian scribes to write determinatives like *za* before the name. This was a way of classifying the name of a substance in writing, although the determinative is not thought to have been spoken. In 1936 the Oxford archeologist R. Campbell Thompson compiled a list of 152 Sumerian stones and objects in stones. Among these were the Sumerian *za bil*, literally meaning "burning stone" and interpreted as *pyrite*.[1]

Theophrastus (c. 372–287 BCE) described a mineral, *spinos* (σπινος), which may have included pyrite, in his book *On Stones*, written about 325 BCE. Theophrastus's work is noted for its modern scientific rationality: it is free from fable and magic and remained one of the most rational mineralogy texts for almost 2,000 years. The reason that *spinos* is thought to contain pyrite is because Theophrastus writes that fire is kindled from it. Some authorities think that *spinos* may have been an early term for the Thracian stone, probably an asphaltic lignite that is mentioned by several ancient authors, including Nicander, Dioscorides, and Pliny. Others think that it may have been a pyritiferous, bituminous shale. It would be strange if Theophrastus did not include pyrite in his work in view of its importance for fire-lighting. It is possible that, since the version of *On Stones* that has come down to us is unfinished and fragmentary, Theophrastus may have described pyrite in other, now lost chapters. Theophrastus does describe lapis lazuli as *sappheiros* to describe a variety that contains specks of golden pyrite and is used as a semiprecious stone and *kyanis,* the pyrite-free variety, which was the source of the expensive solid blue pigment so prized by ancient artists. Theophrastus noted that *sappheiros* is spotted with gold, and several ancient authors followed this early mistaken identification of gold for pyrite.[2]

In *De materia medica*, a five-volume work, Greek physician Pedanius Dioscorides (c. 40–90 CE) described the mineral pyrite mainly in the context of its oxidation product (probably alum) as a useful additive in certain medicines.[3] This volume has been described as the foundation of Arabic pharmacology and, as such, provided a key link between the mineralogy of the classical world and that of Renaissance Europe.

In his *Natural History*, the elder Pliny (c. 80 CE) mentions some 20 Greek writers as his authorities on minerals, but only the fragmentary treatise by Theophrastus *On Stones* and the two poems by Nicander of Colophon (2nd century BCE) have been preserved. He is probably responsible for the first use of the mineral name, *pyrites*, that has come down to us directly. Pliny was a great collector of ephemerae, and he notes initially that some people use *pyrites* to describe a type of millstone—this obviously not referring to an iron sulfide. However, he goes on to describe other sorts of pyrites (which he describes as "live pyrite") and gives a surprisingly clear description of the mineral, noting that it had a golden color and was used in strike-a-lights. Perhaps the reason for the unusual clarity in this otherwise rambling and gossipy discourse is that it was mostly derived from a lost treatise on stones by Xenocrates of Ephesus, a 1st-century CE author of the standard Hellenistic era lapidary, the now lost *Lithognomon*.[4]

Pliny refers to *pyriten, pyrites,* and *pyritarum*. This seems a bit strange grammatically for us schoolboy Latin scholars. However, it appears that these are common endings for Latin nouns derived from the Greek: *pyriten* is the accusative plural, *pyrites* is the nominative plural, and *pyritarum* is the genitive plural. Note that all Pliny's pyrites are plurals and seem to have been used to describe a group of minerals. Interestingly, the nominative singular (which Pliny never uses) would be *pyrite*. The Greek root of the word adds weight to the idea that pyrite was first defined in a currently lost manuscript by a Greek author

Marcasite and the Arab Interlude

In modern mineralogy, *marcasite* is specifically reserved for a dimorph of pyrite: that is, it has the same chemical composition as pyrite but a different crystal structure. In jewelry, *marcasite* refers to pyrite: marcasite itself is far too brittle and unstable to be used in jewelry. Pyrite was used in jewelry by the ancient Greeks, Romans, and Incas. Its popularity in England began in the 18th century where it was substituted for the cut steel beads in Georgian times. It really became popular as *marcasite* in Victorian times after the death of the queen's husband; cut pyrite set in silver and marketed as *marcasite* was considered appropriate for decorating the dark and somber fashions of that period (Figure 3.2). Why the jewelry trade chose to call pyrite *marcasite* originates with the Arab philosopher-scientists of the early medieval period and the invention of alchemy. The romance and mystery of the Middle East affects European mores even today and has led to *marcasite* being the more widespread and recognizable term in the public consciousness because of its association with pyrite jewelry. The giveaway is that the jewelry trade commonly (and I choose the word carefully) pronounces *marcasite* "marcaseet."

This modern misappropriation of the name *marcasite* is not a new phenomenon. Marcasite has had a somewhat rocky history: it was also used as a term for antimony and bismuth in the Middle Ages. The Chinese may have recognized marcasite, and *Te sheng bai shi* (特生白石) or *special white ore* may have been used to differentiate marcasite from pyrite. But, as in the case of *wasserkies* in Europe, it is possible that the term was also used for arsenopyrite.

FIGURE 3.2. Typical Victorian marcasite brooch showing polished and faceted pyrites set in sterling silver (see color plate). Courtesy of www.antique revisions. com.

Marcasite is generally agreed to have entered western European literature from the Arabic *mārqashītā*. The puzzle is why the Arabic philosopher-scientists used this term instead of, or sometimes in addition to, pyrites, which they transliterated as *būrītis*. After all, much of the early Arabic work was based on translations of ancient Greek works, and it would seem logical that pyrite would be the preferred name. For most of written history *marcasite* has been used as a synonym for *pyrite*, even to the extent of referring to the same group of sulfide minerals. From our point of view, the question is: Why did the Arab authorities have a different name for pyrite? The Greek root of pyrite clearly refers to its fire-lighting properties, but what is the etymology of *mārqashītā*?

There is a general problem in etymological studies: we can never really know what an ancient person meant by a particular word at a particular time. The cultural markers that delineate, at least approximately, word meanings in the present day are less clear in the past and are sometimes absent. Furthermore, the plasticity of meaning is familiar to us today as words appear to morph into alternate meanings with time. In the case of ancient literature, this problem is compounded by the lack of any prescribed, contemporary standard language. Thus the scribes wrote down what they heard with the best phonetic symbols and pictographs available. In Arabic and its progenitor languages this is made more difficult by the replacement of vowels by dots or their entire absence, making the pronunciation of words difficult to ascertain. We may never know what actual sound an ancient scribe was reproducing in his script. So when we begin searching for the origins of the word *marcasite* and why the Arabic-speaking peoples appear to have preferred this to *pyrite*, we enter a world of speculation based, it must be admitted, on some of the most rigorous scholarship.

The oldest civilization in the Middle East was the Sumerian in the 4th and 3rd-millennium BCE. The Sumerian cuneiform language is unrelated to any other language. However, Akkadian, which succeeded Sumerian in importance in the 3rd-millennium BCE, used the Sumerian cuneiform script. Akkadian gave rise to both Assyrian and Babylonian in the 2nd millennium BCE. A late Babylonian tablet was discovered in Uruk at the beginning of the 20th century that included a word list of Akkadian equivalents of the Sumerian, all introduced by the determinative *za*, meaning *stone*.[5] Among the list of 152 Sumerian and Akkadian stones was an Akkadian mineral *za marhaši*, which has been translated as *marcasite*. *Marhaši* was in fact the name of a 3rd millennium BCE state situated east of Elam in the Zagros mountains of modern western Iran.[6] So the complete Akkadian inscription *za marhaši* means literally *Marhaši stone*. Marhaši became part of the Akkadian Empire under Sargon the Great, who lived sometime between 2334 and 2215 BCE.

It is interesting that the Sumerians appear to have had only one word for this mineral, *bil*, which relates to the early use of pyrite in fire-lighting. However, the later Akkadians had two: *pindar*, relating to fire-lighting, and *marhaši*, describing where it came from. The use of geographical terms to describe minerals and rocks continues well into the modern era, especially when samples from different sources might have different qualities, appearances, compositions, and properties. Today the same technique is used in the building and decorative stone industry. In the United Kingdom, commercial stones are given names that help sell the product and do not necessarily describe its source or even composition. For example, the finer-grained variety of the beautiful red Rapakivi from southwest Finland is marketed in the United Kingdom as Balmoral granite, with an allusion to the royal estate in Scotland. So I suppose that *Marhaši stone* had a better cachet than run-of-the-mill *pyrite* to the Akkadians, rather like the modern jeweler's use of the word *marcasite* for *pyrite*. Certainly, there were few sources of pyrite in the fertile plains that occupied the heart of the Akkadian civilization; the metals were mined from mineral deposits in the mountains, especially the Zagros.

The origin of the word *marcasite* and why the Arabs and their predecessors used this term instead of—or at least in addition to *pyrite*—remains a matter of speculation. The Marhashi derivation idea is attractive. I do not speak Arabic, and I rely on the Latinized transcriptions of the script. My ignorance is so great that I really did not know how مارقشيتا (*mārqashītā*) is actually pronounced. I went along to see my friend Kurda Saied-al-Berezanchi, a classically trained modern Arabic speaker. I asked her how she pronounced the Arabic word *mārqashītā*, by pointing at the Arabic script like the simplest kind of English tourist. To my surprise she softens the *t* so that it sounds much more like *marquashīshā* and much closer than I supposed to Marhashi. Indeed, *marquashīshā* is a widespread variant of *mārqshītā* in older medical prescriptions.

The oldest written use of the term *marqashītha* in Arabic may be the comment inserted on *būrītis* or pyrite in the Arabic translation of *On Sulfurs* by Zosimos of Panopolis.[7] Zosimos was an important figure in early alchemy since he was the author of some of the oldest known books on the subject from the 4th century CE. Zosimos was born in the south of Egypt but was steeped in classical Greek culture and learning. His text survives in fragments in an Arabic translation said to have been written in 659 CE but which was probably actually from the 9th century.

Jabir (Abu Musa Jabir Ibn Hayyan, also known as Geber) is a mysterious figure from the 9th century CE. He has been identified with an historical, probably Persian, alchemist who lived 721–815 CE and worked at the court of Caliph. However, the first reference to Jabir was in a catalogue prepared by Ibn al-Nadim, the librarian of Baghdad in 988 CE, almost 200 years after his death. Works attributed to Jabir number over 3,000, of which several hundred are probably Arabic in origin. These Arabic works constitute an encyclopedic collection that summarizes what medieval Islam knew about science. It is credited to the 8th-century Jabir but Paul Kraus, the foremost Jabir scholar, thought it to be primarily the *nom-de-plume* of a group of Ismaili propagandists writing in the 9th and 10th centuries.[8] Some authorities neatly distinguish between the Arabic writings of Jabir and the Latin writings of Geber, a pseudonym used by a 13th-century European alchemist. Because of the uncertainty of the provenance of much of Jabir's work, it is difficult to discuss his knowledge of pyrite and marcasite. Even Jabir's *Book of Stones according to the Opinion of Balinas*[9] (*Kitab al-Ahjar'alá ra'y Balīnās*) appears to be a 10th-century work that, however, fails to mention marcasite or pyrite and is basically esoteric ramblings. It does seem, however, that Jabir used *marcasite* rather than *pyrite* as a name for iron sulfide.

The Jabir records reveal that pyrite in early medieval Persia had developed a series of new technological uses related to the cutting edge of contemporary technology. The Jabir authors record a recipe for ink based on mixing iron sulfate or vitriol, derived from pyrite as described in Chapter 2, with tannic acid extracted by fermentation of oak galls. The ink was bound with gum Arabic and produced a smooth, rich, purplish-black permanent ink that was the favored medium of Roman scribes and Renaissance artists alike. The Arabic scientists did not invent iron gall ink: Pliny the Elder provides possibly the first account of it in the 1st century CE. However, Pliny's description was, as usual, vague and Jabir's recipe is far more precise. Iron gall ink was used for over 1,500 years until it was replaced by chemical inks in the 20th century. The purple-black color that developed in the reaction between iron and tannic acid provided the basis for the gall test for the presence of iron. This test was the basis for the original demonstration that pyrite is an iron mineral in the 17th century CE, as described later. Jabir is credited with a book of forty-six recipes for colored glass, two of which include marcasite as an ingredient.[10] Glassmaking had been around since 3500 BCE and had been industrialized by the Romans. However, once again the Jabir volume provides one of the first listings of clear recipes for glassmaking, a technology shrouded in secrecy for much of its history. The clarity of Jabir's recipes compared with earlier writers contrasts markedly with his stated aim that the purpose of his work was to baffle unenlightened readers. It adds to the idea that Jabir's works are not those of a single author.

Aristotle's *Book of Stones* became the preeminent source of information about minerals in the early Middle Ages and was much sought after by medieval scholars. It was originally compiled sometime before the middle of the 9th century by a Syrian acquainted with Persian and Greek traditions.[11] This work was written in Syriac and translated into Arabic in the 9th century. Syriac, a variety of Aramaic, was the literary language of the 4th to 8th centuries CE. The Syriac for *marcasite* is *mqshytā*. The Aristotle name—even though this did not refer to the original Greek Aristotle—meant that the book was treated with undeserved reverence by later Arabic and European scholars. Many of the mineral names are considered to be of Persian origin, and that may be true of *mqshytā* too. Translated into Arabic, it is the oldest known Arabic authority on minerals. It has been translated and rewritten many times in a variety of languages, so the content varies widely between versions. One version lists seventy-two minerals and includes marcasite as number 24 (see endnote 11). An interesting aspect of the marcasite description in Aristotle's *Book of Stones* is its similarity to descriptions of pyrites in the European literature of the late Middle Ages. Thus marcasite is described as occurring in a number of varieties, including gold, silver, and copper species. It was therefore used in the same way as *kies* was in later times.

Hunayn ibn Ishaq (809–873 CE) translated many Greek scientific and medical works into Syriac and Arabic. Hunayn's translation of Dioscorides includes the direct translation of the Greek *pyrites lithos* as *al-hajaru allādhī yuqālu lahū būrītis*, that is, "the stone called pyrite." But Hunayn often added comments later to the margins of his manuscripts to better explain the text, as was the case in the Zosimos text. Hunayn felt that his readers would not be familiar with the mineral *būrītis* or pyrite and added *wa-huwa al-marqashīthā* (that is marcasite; see endnote 7). The conclusion is that the name *marcasite* was first used in Arabic before the 9th century CE, since Hunayn refers to pyrite being better known to his Arabic readers as marcasite.

By the turn of the 11th century CE, the great Islamic polymath Abū Rayhān al-Bīrūni introduced Indian and, thereby, Chinese mineralogy to Arabic. Al-Biruni listed over 100 minerals in his collected information on precious stones.[12] He described marcasite in some detail:

> Glitter in the stone is brought about by golden marcasite. Mineralogists call golden marcasite, *turunjah*, since its paleness is like that of the citron. Marcasite comprises several varieties: golden-yellow, silvery-white, brassy-red, and pyrite-black. The polishers use the golden-yellow kind.[13]

Al-Biruni also uses the name *pyrite*:

> The mineral from these mines [in the region of Bart] is high quality, firm and clean. Besides, there are gold and silver, white and golden-colored pyrites, and other allied stones.... This city [Baqim] also has the Hinduwan mine, while marcasite is to be found more abundantly in Syria.

Al-Razi (Muhammad ibn Zakariya al-Razi, Latinized as Rhazes) was a Persian scholar who wrote mainly in Arabic and lived 865–925 CE. He published some 200 books, mainly in medicine. In his *Book of Secrets* he organized minerals into six divisions, including spirits, bodies, stones, vitriols, borates, and salts. Pyrite and marcasite were listed as stones. Al-Razi was very much concerned with how minerals such as pyrite form—that is, how they developed their distinctive properties of shape, color, luster, malleability, and density. He proposed a theory of transmutation whereby several agents, called variously medicines, tinctures, and elixirs, induced changes into an original metal.

Al-Razi observed that pyrite—as marcasite—in contrast to many common rock-forming minerals and even gemstones released sulfur on heating, leaving behind a dull, amorphous lump of black, metal-rich stone. He concluded that the reaction between an elixir like sulfur and a stone transmuted the metal-rich stone into lustrous, golden crystals of pyrite. In other words, the elixir caused the transmutation of the properties of the stone, such as color, luster, and shape. It was not a great leap of the imagination to conclude that all substances could be transmuted and become more perfect in form. Ultimately gold could be produced. The experimentalists had some evidence for this since pyrite often contains significant amounts of gold and this could be extracted by cupellation with mercury. In other words, pyrite could be changed partly to gold. We can see how these observations and writings of these Arab scientists gave rise to both the science of chemistry and the dead-end of alchemy.

Rhazes, as Al-Razi became known in Europe, was particularly significant; his *Book of Secrets* was translated into Latin and had a substantial effect on later European thinking about minerals. It is really a list of laboratory equipment and materials rather than an in-depth probe into how transmutation occurred. In modern terms it constitutes the methods and results sections of a scientific study with relatively limited discussion. I must admit that Rhazes is a bit of a hero of mine since he showed us how to probe nature experimentally. In particular, he demonstrated if you put the same ingredients into a pot and do the same things to it, you get the same—that is, repeatable—results. In this sense his work was an early example of one of the fundamental laws of modern science, that of reproducibility. Rhazes' careful, detailed, and relatively exact experimental recipes enabled later workers to repeat his experiments and build on his results. Of course, it also gave rise to wide-eyed speculation by the later alchemists. Rhazes' *Book of Secrets* is clearly one of the primary sources for the most popular alchemical text of the Middle Ages, the *Summa perfectionis magisterii* published under the Geber pseudonym.

Al-Razi's successor was Avicenna, which is the Latinized name of Ibn Sina, who translated and commentated on Aristotle and wrote a mineralogic section in his scientific and philosophical encyclopedia or *Book of Healing of the Soul* or *Kitabal-shifa,* which was published in 1027 CE. In the mineralogic section it has an entry for ﻣﺎﺭﻗﺸﻴﺘﺎ *mārqashītā.* Part of what Ibn Sina writes about *mārqashītā* closely follows the text of Aristotle's *Book of Stones.*

The *Kitabal-shifa* was translated into Latin by Alfred of Shareshill in 1190 CE as *De Congelatione et Conglutatione Lapidum.* The title was derived from Avicenna's thesis that minerals may be formed either by solidification from a liquid or by aggregation of solids. The book was also known as the *Liber de Mineralibus* and was ascribed to Aristotle for many years. In this setting Avicenna's thesis on how minerals like pyrite form resonated down through the ages and, as discussed further on, can be most clearly traced in the paradigmic work of the German scientist Georgius Agricola some 500 years later. Avicenna is particularly renowned in chemistry for his attack on alchemy. He stated clearly that chemical transmutation of metals did not occur. In a neat twist, Alfred of Shareshill's translation of the mineralogic part of Avicenna's great medical treatise was originally bound in with a translation of Aristotle's *Meteorologia* and was accepted as the missing last part of this work.[14] So Avicenna's attack on alchemy came to Europe with Aristotle's imprimatur.

The Arab interlude left the world in general and Europe in particular with two names for the minerals of the pyrite group: pyrite and marcasite. There was essentially no difference between them, and both refer to a group of minerals that included varieties with valuable constituents, such as gold, silver, copper, and zinc. Possibly because of the Arabic origin of the word *alchemy,* medieval alchemists and their followers may have found the name *marcasite* more exotic than *pyrite* and preferred it in their secret and often magical recipes. *Pyrites* may have been a more ordinary name for the same material.

It is significant that for this whole early medieval period the Arab philosopher-scientists did not realize that iron disulfide occurred in two different crystalline forms, which we refer to today as marcasite and pyrite. It appears that they used *marcasite* as a simple synonym for *pyrite* and, like *pyrites,* marcasite was essentially a mixture of metal sulfides. As I mentioned before, in trying to trace the contribution that pyrite has made to science and culture, the primary problem is to understand what ancient scientists meant by various mineral names. In the key period between the fall of the Roman Empire and the early Renaissance, Arabic philosopher-scientists referred to pyrite as marcasite. In this form, pyrite played a significant role in the transmission of classical Greek and Roman scientific ideas to medieval Europe. The contribution of the great Arabic philosopher-scientists was not merely limited to translation of otherwise lost classical texts but was key to the development of the embryonic sciences of chemistry, geology, and medicine during the early medieval period.

Late Medieval Confusion

The 13th century saw the establishment of universities in western Europe, the rise of the mendicant Dominican and Franciscan orders, and the theology and philosophy of the Schoolmen. One trigger for this intellectual expansion was the rediscovery of the classical Latin and Greek texts, albeit through the prism of Islamic science of the 7th to 10th centuries. This was accompanied by the publication of a number of didactic books intended to compile all knowledge of things or nature. This was the century of the encyclopedists.

The interests of the intellectual elite of this period were completely different from our present science. Their aim was to try and fit observations of natural things into classical philosophical theories and to reconcile these with contemporary Christianity. This is why they are called the Schoolmen. They spurned new observations or experiments as unnecessary, since their purpose was not to increase knowledge per se but to enhance understanding. Preeminent among the ancient authors whose philosophy they followed was Aristotle. Since it was believed that Aristotle did most of his teaching walking about in the Lyceum in Athens, these medieval followers of Aristotle were called Peripatetics, and they were described as the Peripatetic School.

Albertus Magnus was a Dominican monk and is widely regarded as the greatest intellect of his era. He was canonized in 1931 and, in 1941, Pope Pius XII declared him the patron saint of scientists. Albertus lived about eighty-seven years in apparently robust health throughout his life. Even taking into account his long and healthy life, his writings are enormous and constituted some thirty-eight volumes when collected together in 1899. They also cover virtually all aspects of medieval science, arts, humanities, philosophy, and theology. But what Albertus is particularly renowned for is the translations and commentaries of all the extant works of Aristotle.

He brought Aristotle back into the western European consciousness, and this formed the cornerstone of science through the medieval period to the Renaissance. Albertus's *Book of Minerals* is his own work. Arabic tradition claimed that Aristotle's *Book of Stones* listed 700 stones, so you can see why Albertus was so keen to get his hands on it. In fact, the existing text lists between 80 and 100 stones. However, Albertus had access to only brief excerpts from Aristotle's *Book of Stones* and could not locate a copy of Aristotle's *Mineralia*. He therefore drew up his own plan for what an Aristotelian mineralogy should look like. Book II of his *Book of Minerals*, concerning precious stones, is largely taken from the lapidary of Arnold of Saxony, together with some readings from Thomas of Cantimpre and Bartholomew of England.[15]

The significance of the Peripatetic approach to pyrite and other minerals is that the Peripatetics were not interested in describing minerals themselves or their properties but were concerned how the minerals fit in to the Peripatetic philosophy and, of course, how this could be reconciled with the contemporary Christian faith. This should not be strange to us today. The prime purpose of the €70 billion scientific research budget of the European Union is not to progress science but to further facilitate international cooperation within the Union, at least according to the past European Commissioner for Research, Science and Technology, Edith Cresson.

The idea of the Schoolmen was that data gained by direct observation of minerals are of concrete particulars but are often confused and difficult to understand. The Peripatetic approach concerned itself with analyzing extant data to make things understandable by explaining their *Causes*. *Causes* in this sense is capitalized since the Peripatetics were referring to the Four Causes of Aristotle: *material, efficient, formal*, and *final*. This meant that something as basic as a mineral name was not particularly important to them. It also means that tracing the development of the understanding of the properties and nature of pyrite during this period is difficult.

The confusion of the names used for pyrite in ancient through medieval texts was illustrated by the work of Arnold of Saxony, the 13th-century encyclopedist who provided Albertus Magnus with his list of minerals.[16] Arnold himself wrote *virites* in Volume 3 of this work and *pyrette* or *pirette* in various versions of the manuscript to Volume 4. At least this is how the copyist in some manuscripts transcribed it. It seems that *virites* was simply an artist's error when illuminating the first letter of the text beginning with *pirites.* Arnold had more or less copied the section of his encyclopedia on minerals almost entirely from the *Book of Stones,* a famous medieval lapidary written by Marbode, Bishop of Rennes in 1096 that listed sixty stones.[17] The error was compounded by Albertus Magnus in the part of the *Book of Minerals* he copied from Arnold. Albertus wrote *uirites,* which is understandable considering that the Latin *u* and *v* were interchangeable. He also listed a different mineral, *perithes,* which is also pyrite, and he copied this from his colleague Thomas of Cantimpre. Dorothy Wyckoff [18] identified *marchasita, adestrum, chyselectrum, epistrites, topasion,* and possibly *andromata* as minerals listed by Albertus that were probably varieties of pyrite. This is significant because Wyckoff was a distinguished mineralogist, whereas most other commentators and translators were humanists. Meanwhile, Vincent of Beauvais and Bartholomew of England wrote *pyrites.* The great *Liber Aggregationis,*[19] a compendium of some eighty medieval manuscripts that went into some 350 editions and was a bestseller in the 14th and 15th centuries, includes *fendanius, urites,* and *puricem apix* as apparent synonyms of *pyrite.*

The problem of the various names that may or may not have been used to describe pyrite is not only a feature of European scientific literature of the time. As mentioned in Chapter 2, the ancient Chinese called pyrite by various names that were related to what the mineral—especially the sulfur product—was used for or where the mineral came from.

Following Islamic tradition, the name *marcasite* was often used as an alternative name for *pyrite* in European literature. Albertus Magnus discussed a mineral he called *marchasita* in some detail. Dorothy Wyckoff described *marchasita* as an alchemical name for metallic sulfides, including pyrite and marcasite but also other sulfides such as stibnite, the antimony sulfide. Albertus's interest in *marchasita* was because of its use in amalgamation for the separation of small amounts of gold that often occurs in pyrite. The crushed pyrite is mixed with mercury and the gold amalgamates with the mercury to produce a gold amalgam. The amalgam is heated and the mercury driven off as a vapor, leaving the gold behind. The alchemical interest is obvious because the pyrite appears to have changed or transmuted to gold. Albertus thought that the golden varieties of *marchasita* were substances that had not yet achieved a perfect golden form but were on their way. Albertus reported that *marchasita* could be found in a variety of forms, probably related to the degree of perfection attained by the mineral at the time. We recognize these forms today as different sulfide minerals.

FIGURE 3.3. Pyrite flower: a broken surface of a 5-cm-diameter pyrite nodule show-ing the radiating pyrite crystals (see color plate).

The marcasite appellation is still used to describe pyrite in some quar-ters. Thus, as noted in Chapter 2, radiating pyrite nodules (Figure 3.3) are often called marcasite because of the crystal habit. More often than not they turn out to be pyrite.

Renaissance Revisions

Georgius Agricola is best known for his great work *De re metallica*, which described mining technology in the late Middle Ages. Agricola's work laid the foundation for modern metallurgy, mining, geology, and mineralogy. He was based in central Europe, which was the main source for metals such as copper, zinc, lead, gold, and silver in the Middle Ages, and these met-als were mainly mined from pyritiferous ores or *kies*. Agricola's approach was therefore heavily colored by these central European ores, and German mining and metallurgical engineers were exported to mines and smelters all over Europe. They took their experiences with the massive sulfide ores with them. Their expertise dominated European and ultimately global mining and metallurgy until the 20th century, when mines from overseas colonies began to compete with and ultimately overtake European producers. For example, German miners went to Sweden in the 15th century, and their successors developed the original flotation techniques in the beginning of the 20th century whereby valuable metals, such as copper, zinc, lead, silver, and gold, could be separated economically from massive sulfide, usually pyrite-dominated, ores. This process is still the basis for extractive metallurgy of these ores worldwide today. The importance of the work of Agricola and his German successors to our understanding of pyrite and pyrite-rich ores should not, therefore, be underestimated.

Agricola knew of Dioscorides' work and included a discussion about this in his first book *Bermanus: A Treatise on Mineralogy*[20] published in 1530. This is an unusual mineralogy text: minerals are described by means of a dialogue between three people, two physicians and Bermanus, a miner, as they wander through a mine in Saxony. They discuss the correlation between what they see and the mineralogic descriptions from classical sources. One of the physicians remarks, "Perhaps marcasite is the same as pyrite?" In Book 10 of his later, more orthodox text *De natura fossilum*,[21] Agricola describes *pyrite* as a generic term for a group of minerals. Again, keep in mind that Agricola used the term *pyritus* and modern translators have often rendered this as *pyrite*, whereas it is better translated as *pyrites* as discussed previously.

Martin Rulandt was a German physician and alchemist whose *Lexicon alchemiae* was published in 1612. The purpose of this work was to clarify alchemical terms and provide definitions rather like an alchemical dictionary. Rulandt was very clear about pyrites and marcasite, and his entry for "pyrites" starts off with:

> Pyrites and Marcasite are the same, for what the Romans and Greeks called Pyrites the Arabs term Marcasite and Black Zeg.... All that the Greeks have written concerning Pyrites the Arabs ascribe to Marcasite in their own language. It is called Pyrites because fire is often struck from it.[22]

Rulandt lists some seventy-one different types of pyrites, including silver, golden, and iron-colored varieties, and notes that the golden type has a higher proportion of sulfur. Like Agricola, he had little time for Albertus Magnus: in this entry he stated that Albertus was wrong to write that

metals cannot be smelted from pyrites or marcasite and doubts whether
Albertus had actually read Avicenna and Dioscorides. It appears that by
the turn of the 16th century there was some clarity developing in central
Europe about pyrite and marcasite—even though it was still wrapped in
the nonsensical ramblings of contemporary alchemy.

Johann Friedrich Henckel published a 1,000-page volume in 1725 titled
Pyritologia. My copy of this book is a squat volume just 17 cm tall and
10 cm wide. The book size is called Foolscap 8vo by bookmakers, and it
is bound in brown leather with a ribbed spine. A label stuck on the spine
declares the book to be *Henckels Kies Historie*. The book is printed in what
I would call German Gothic (Figure 3.4) but which I am told is actually a
variant called *Fraktur,* which was popular in Germany after the mid-16th
century. Whatever you call it, it is difficult to read. The text is bookended
by splendid small ink drawings. The frontispiece (Figure 3.4) shows miners at work, and the twelve plates at the end are drawings of pyrite. One is
shown in Figure 4.2 in Chapter 4 of this volume. Henckel was the director of the mines in Freiberg in Saxony and the leading mineralogist of
the first half of the 18th century. His book was translated into French and
English and went into several editions. The English translation of 1757
gives the title as *Pyritologia or, a History of Pyrites, the Principal Body in the
Mineral Kingdom.* This is interesting since the actual German subtitle is
Kieshistoria (as shown in Figure 3.4), which would have been more accurately translated as "the history of sulfide ores."

The title page goes on to detail the content, which was translated into
English as:

> In which are considered its names, species, beds, and origin; its
> iron, copper, unmetallic earth, sulphur, arsenic, silver, gold, origi-
> nal particles, vitriol, and use in smelting. The whole compiled from
> a collection of samples; from visiting mines; from an intercourse
> and correspondence with naturalists and miners; but chiefly from a
> course of chymical enquiries. With a preface, containing an account
> of the advantages arising from mine-works in general and, in par-
> ticular, from those of Saxony.

Henckel's volume consists of sixteen chapters, and its organization is
similar to that of a modern-day text. It includes chapters on the history of
pyrites, the types (species) of pyrites, iron in pyrite, sulfur in pyrite, and
even its oxidation products ("Of the Vitriol in Pyrites"). Where it might

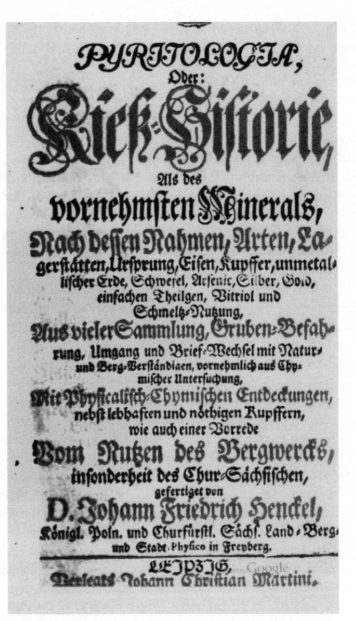

FIGURE 3.4. The title page of Henckel's great 1725 book, *Pyritologia*.

diverge is in the chapters on the copper, arsenic, gold, and silver in pyrite, which reflect Henckel's use of the term *pyrites* to include a number of metallic sulfide minerals.

The Iron Mystery

It seems extraordinary that the ancients apparently did not know that pyrite contains iron as its other major element. Pliny, for example, used some twenty Greek authorities (of which only Theophrastus and Nicodemus have come down to us, as mentioned earlier) but failed to report that pyrite is an iron mineral. However, the Aristotelian view of the world dominated through to the Renaissance, so the question of whether or not pyrite contained iron was not relevant inasmuch as the mineral was not used as a source of iron.

The idea that there were four elements—earth, water, air, and fire—has extremely ancient origins. To these were sometimes added a fifth element called *quintessence* (from *quint* or *fifth*) or *Aether*. The oldest reference to the four or five elements is in the Babylonian *Enûma Eliš*, a creation myth from the 18th to 16th centuries BCE. Here the five elements are sea, earth, sky, fire, and wind. Strictly speaking these substances were not called elements until Plato termed them *stoicheion*, literally the smallest divisions of a sundial. Empedocles of Agrigentum (490–430 BCE) is generally credited with being the first philosopher to insist that there were just four elements. Empedocles actually called these *roots*. Aristotle (384–322 BCE) reverted to a fifth element or quintessence since the stars were unchanging and thus could not be made of the four mutable elements. Aristotle's views on the nature of materials provided the basic model for the next millennium.[23] All substances were mixtures of these four elements, and their properties were derived from the proportion of each element the material contained. In the case of metals, many could be melted with the application of heat. They were thus assumed to contain varying amounts of water, which was released by heating and produced the fluidity observed in the molten material. This water was mixed with earth. Earth was essentially dry and the main component of rock, or matter that did not appear to melt.

In 1546 Agricola[24] classified pyrites that contain metals such as copper, silver, gold, tin, or lead as a compound mineral (i.e., *mistae* or literally *mixed minerals*) composed of stone and metal. Ordinary pyrite was an ore of sulfur (*lapis fissilis*), which was a mixture of solidified fluid (*succus concretus*) and stone. The solidified fluid was basically water in which were dissolved various amounts of earth. This idea harks back to Avicenna's thesis that minerals form by solidification or by aggregation. And it is worth remembering that, as noted earlier, Avicenna's work on minerals was ascribed to Aristotle at the time and so had an impeccable provenance.

Pyrite itself does not melt under any conditions that medieval metallurgists or their ancient predecessors could achieve. It breaks down on heating with the evolution of sulfur and sulfur oxide gases, leaving behind a heavy, iron-rich slag. Thus the idea that it was a mixture of sulfur (the solidified fluid) and stone is perhaps not surprising. It did not appear to contain the basic element of water and thus, by definition, did not contain any metal since metals were mixtures of water and earth. The iron-rich slag produced by burning pyrite in a closed furnace with limited air access, as described for sulfur production, can produce magnetite. Magnetite is a black, heavy, iron-rich mineral that is not only magnetic (the ancients knew this as lodestone) but is smeltable with a carbon source, such as wood charcoal, to produce metallic iron. Indeed, even in 2011 the Pyhasalmi mine in central Finland produced over 800,000 tons of pyrite concentrate and sold the slag from pyrite burning to blast furnaces. So we would expect some ancient alchemist, metallurgist, or smith to have done this. However, it does not seem to have reached the writers of ancient texts. Perhaps there is a clue in a closer reading of Agricola's description of pyrites in *De natura fossilum* (see endnote 21): "Sometimes, however, pyrites do not contain any gold, silver, copper or lead and yet it is not a pure stone, but a compound, and consists of stone and a substance which is somewhat metallic, which is a species of its own."

The qualification that pyrite contains a substance "which is somewhat metallic" may be a reference to the observation concerning the nature of the slag that is left after sulfur production. It may well be that, like Homer's blue, it was so obvious that pyrite was an iron mineral that no one thought to write it down. However, Agricola was the leading metallurgist, mineralogist, geologist, and mining engineer of his age, so we can be fairly sure that the fact that pyrite was an iron mineral was unknown up until that time.

A second strand of ancient thought about the nature of matter was provided by the alchemists. Chinese daoists, Egyptian necromancers, and Greek philosophers all appear to have embraced alchemy. Aristotle provided a philosophical background to alchemical thought. Thus matter is constituted by the four elements—earth, fire, water, and air—but behind these four elements was the unstructured *prima materia*. The alchemists were basically searching for this first material from which all other things, including life itself, derive.

Mercury was a key aspect of this *prima materia* because it was a metallic liquid at room temperature. Thus the alchemical idea was that, since metals become liquids under the influence of heat, they must contain varying amounts of mercury. The alchemists considered metals to be combinations of varying proportions of mercury and sulfur. Sulfur is an interesting choice in this context, probably selected on the basis of its sublimation as an earth-rich material. It is also the case that early metals were not pure and likely to contain small amounts of sulfur. Heating them then produced a sulfurous smell, as described by the Franciscan alchemist, Paul of Taranto, in the 13th-century writing under the Geber or Jabir pseudonym.

The consequence of this alchemical thinking is that it was widely accepted that any metal could be changed into another one (transmutation) by altering the proportions of the mercury and sulfur. Thus lead could be changed to gold, for example. This idea was supported somewhat by the extraction of gold by amalgamation with mercury.

By the Middle Ages, therefore, two strands of chemical thought could be distinguished: the pure Aristotelian idea of the four elements and the more mystical alchemical philosophy based on the idea of the *prima materia*. As discussed previously, the alchemical approach probably reached its apogee in the West with the work of Albertus Magnus (1193–1280 CE). It is also not coincidental that the work of the Arabic alchemist Jabir is mainly known through a 12th-century Latin compilation by Pseudo-Geber: the *Summa Perfectionis Magisterii*. This was one of the most influential alchemical books in the late medieval period. It was probably written by Paul of Taranto, the Franciscan alchemist mentioned earlier who may have compiled the work around 1310 CE. The Pseudo-Geber work strongly promoted the idea that metals were mixtures of mercury and sulfur in different proportions. Albertus came into contact with the Arab alchemists through his studies of the writings of Aristotle, which had been passed down to the medieval scientists through Arab authorities. Albertus writes that he had seen the creation of gold by transmutation, a statement that led to widespread ridicule by later workers such as Agricola and Rulandt.

Renaissance Clarification

By the time of Agricola in the 16th century, alchemy was in decline and widely discredited through the activities of charlatans, tricksters, and fairground showmen. Agricola himself took strongly against alchemy and dedicates much of his work to criticizing Albertus Magnus. Agricola took a purer, Aristotelian line.

By contrast, Johann Friedrich Henckel knew in 1725 that pyrite was an iron sulfide. During the intervening years one of the great founders of the science of chemistry, the Irish polymath Robert Boyle, had demonstrated in 1664 that the gall test actually revealed the presence of iron.[25] The confused account of the gall test in the 1st century CE by Pliny the Elder had not specified any particular substance as being detected by the test. All Pliny was concerned about was describing a method of determining the purity of verdigris, or copper acetate, an important medicinal salve to the Romans. Unscrupulous dealers would substitute cheaper green vitriol or iron sulfate for the more expensive verdigris, and this fraud could be detected with the gall test. Boyle showed that the blackening of gall was specific to the presence of iron: chemists then had a recipe for qualitatively proving the presence of iron even though they were still unable at that time to analyze it quantitatively.

In 1725 Henckel cited that the fact that pyrites contain iron was first reported in 1682 by Martin Lister, an English medic who was Queen Anne's physician[26]. Lister had used the gall test to demonstrate that pyrites contain iron. Henckel writes:

> Dr. Martin Lister, was the first, or among the first, who seems to be aware of iron being the capital constituent of the pyrites; or the first who clearly expressed himself in this matter: *Pyrites puris putrus ferri metallum est,* (the pyrites consists entirely of iron) [Henckel's italics]; whereas in another place he says, *unus Angliae pyrites, purim putum metallum est,* (one sort of pyrites consists entirely of iron). I very much doubt, whether he was quite so sure of the truth of his former enunciation, and whether he knew to give it its full latitude, and true extent.

Henckel misquoted Lister. Lister wrote, "unus Angliae pyrites, purim putum ferri metallum est." Henckel's translation is also faulty: it would be more accurately rendered as "one English pyrites is pure, unadulterated iron metal." Henckel obviously found Lister's account somewhat confusing since Lister clearly knew that pyrites contained a large amount of sulfur.[27] It may be that Lister's phrasing was awry and he was just emphasizing the point that pyrites contained iron. Of course he could not know that pyrite contained no other metal since he had no means of quantitative analysis at that time. So his conclusion that it contained "pure, unadulterated iron metal" was an overstatement.

I think Henckel suspected that Lister was not referring specifically to the mineral pyrite since Lister insisted that all the "pyrites" he analyzed were magnetic.[28] There is a magnetic iron sulfide, pyrrhotite, which was known to the German miners as *magnetkies*. Agricola had used *pyrites* as a general term for sulfide ore minerals—basically a Latinized version of the miner's term *kies*. However, this caused some confusion among scholars who had access only to works in Latin and had not worked with miners who were quite clear what they meant by *kies*. Lister was certainly aware of Agricola's work, and it may be that Lister, who was not a mining man, misunderstood even the limited specifics of Agricola's Latinization of *kies*. He appears to have extended the definition of pyrites to any metal ore, including iron oxide and carbonate ore minerals such as hematite, magnetite, and siderite, since he presented a box of these minerals to the Royal Society in 1684. In fairness to Lister, he does refer to the "brassie lumps" of at least one variety of the minerals he called pyrites. However, it is not clear that Lister analyzed these "brassie lumps" specifically, and his reference to the magnetic qualities of the minerals he did analyze raises some doubts. He also refers to "marcasite" as a separate mineral,[28] which may have been what we now describe as pyrite and was distinct from the mixture of metal sulfides that were called *pyrites* or *kies* at that time.

Lister's work is interesting because it underlines the difficulties encountered by scientists of that period in probing the nature of the material world as well as the problems for later commentators in interpreting exactly what they meant in their reports. Lister was most interested in the oxidation products of his pyrites, the vitriols discussed previously, since they were widely used in medicine. These oxidation products of pyrite may have provided Lister with the materials for the gall test, as was the case with Pliny over 1,600 years previously. The difference was that Lister had Boyle's evidence that the gall test was specific for iron. The problem is that Lister had a generous interpretation of what constituted pyrite, insisted that the minerals he analyzed were magnetic, and both misstated and overstated the conclusions of his studies. Even only a generation later, Henckel found Lister's reports ambivalent and in need of clarification. There is no doubt that Johann Friedrich Henckel established the fact that pyrite is an iron sulfide, even though Martin Lister may have discovered this previously. Mind you, it is probable that many miners, smiths, and assayists were aware of this over several centuries prior.

So by 1725 it was known that pyrite was an iron sulfide. Johan Gottskalk Wallerius, professor of chemistry at Uppsala University, certainly knew in 1747 that pyrite consisted of iron and sulfur. He described it as *sulfur ferro mineralisatum* in the first modern mineralogic textbook and described three different varieties: *kies, marcasite,* and *wasserkies*.[29]

Elements and Minerals

We tend to take for granted nowadays that if we buy some salt at one grocer's shop it will have the same chemical composition as that bought at another shop, and that salt has the same makeup whether it comes in a package or as a crystal of halite. However, before the 19th century this idea was not obvious. Minerals appeared to have a range of components depending on where they came from, and, in the ultimate case, alchemists thought they could change or transmute them.

At this stage it is worth taking a little byway to explain the differences between the modern concepts of elements, minerals, compounds, rocks, and stones. These are often mixed up—especially by students—and it is important to define the differences. *Elements* are the simplest substances that cannot be broken down further using chemical methods. Iron and sulfur are elements. Elements combine to form *compounds*, and *minerals* are naturally occurring chemical compounds. Pyrite is a mineral; its synthetic equivalent is the compound iron disulfide. *Rocks* are mixtures of minerals and are classified by their broad mineral and thus chemical compositions. Thus shales and granites are examples of rocks. *Stones* are bits of rock and have no scientific definition.

At the turn of the 19th century there was a great debate about the composition of chemical compounds, including minerals. The question concerned how elements combine to form compounds, and was at the beginning of the understanding about atoms and molecules. Pyrite, with its splendid golden crystals, was at the heart of the research. Progress was encouraged by the 18th-century discoveries of many of the common elements and advances in analytical techniques.

Jeremias Benjamin Richter was a German chemist and an assayer in the department of mines in Silesia. He found that the ratios by weight of the components consumed in a chemical reaction were always the same. He produced a three-volume summary of his work between 1792 and 1794 in which he proposed the Law of Definite Proportions, which states that chemical compounds and minerals always contain the same proportion of elements by mass.[30] Unfortunately Richter's writing style is impenetrable, and the Law of Definite Proportions is generally ascribed nowadays to Joseph Louis Proust, a French chemist who proposed it 1806. He based this on some ten years of analytical work, the earliest part of which included analyses of pyrite. Proust analyzed two iron sulfides: FeS, which he synthesized in his laboratory, and natural pyrite.[31] He found that pyrite is composed of 90 parts sulfur to 100 parts iron. Recalculating Proust's results, this would mean that pyrite contains 52.6 weight %Fe and 47.4 weight %S. Atomic weights were not known at that time, and Proust was not concerned with providing a formula for pyrite. He wanted to ensure the ratios of these results were reproducible within the analytical uncertainties of the time—that is, that all pyrite crystals had this composition. As we have seen with Henckel and Agricola, pyrite had been regarded by previous scientists as a group of minerals with more or less infinitely variable compositions.

In 1804 Charles Hatchett, a mineral chemist at the British Museum, revisited pyrite analyses[32] and found that the average composition of five different crystal forms of natural pyrite was 53.24 weight %S and 46.75 weight %Fe. This compares favorably with the modern composition of 53.45 weight %S and 46.55 weight %Fe for FeS_2.

Proust's and Hatchett's results are quite remarkable for anyone who has tried to analyze pyrite by conventional chemical methods even today. It is notoriously difficult to get precision in wet chemical methods for sulfur analyses even if the Fe analysis is more dependable. The problem with percentages is that they have to add up to 100. Therefore, if the analysis of one component is in error, the percentage concentrations of all the other components are also wrong. The precision of sulfur analyses can be improved by doing lots of them: one or two analyses will be highly erratic, but take them to an analyst who has been doing these for years and you get very precise results.

Hatchett went on to show that Proust's synthetic iron sulfide was equivalent to the mineral then known as magnetic pyrite or pyrrhotite today. This has a different composition from pyrite. Hatchett helped confirm Proust's thesis that minerals had constant compositions. Jöns Jacob Berzelius, the great Swedish chemist sometimes known as the father of chemistry, analyzed hundreds of compounds and confirmed Proust's law in 1819.

John Dalton was a schoolteacher in northern England who proposed modern atomic theory in 1808.[33] The difference between the modern atomic theory of Dalton and that of the early Greek philosophers such as Democritus is that Dalton proposed that chemical combination occurs between particles, or atoms, of different weights. This meant that minerals like pyrite should have a constant composition since they are made of simple combinations of whole numbers of atoms of iron and sulfur. Dalton calculated the relative weights of elements, including sulfur, in some of his earliest works. Proust's Law of Definite Proportions was important in confirming Dalton's atomic theory. Richter had introduced the term *stoichiometry* to describe the combination of elements in whole numbers. It is not a great leap from these data to the idea that the whole number combinations of the elements mean that the components of compounds must be distinct particles.

Berzelius introduced the modern system of chemical notation in 1811 in which each element is given a simple abbreviation, such as Fe for iron and S for sulfur.[34] In 1828 he produced a table of some 2,000 compounds in response, he wrote, to English critics who thought that his theories lacked data to support them.[35] In this compilation we read for the first time his representation of the composition of pyrite as FeS^2. His method used superscripts to denote the number of atoms on the basic formula. We now write this as FeS_2, of course.

The combined work of these early chemists showed that there were twice as many sulfur atoms as iron atoms in pyrite. Berzelius was fascinated by the fact that the same combination of elements could give different minerals. In 1820 he showed that marcasite had the same composition as pyrite, and therefore marcasite is also FeS_2.[36]

Berthollet Redux

Berzelius was fully aware that the splendid natural crystals of pyrite he analyzed contained impurities. Indeed, in the 1820 paper he notes manganese and silica in his pyrite. The problem is that all minerals contain impurities. In fact, all chemical compounds are impure—it is the level of impurities that is interesting. For example, if you buy some analytical-grade chemicals from a dealer, the impurities will be listed, usually at parts per million levels. This tells you how much impurity there is in 1 million parts of the compound. Enormous efforts have been made to obtain pure compounds, probably not least in the computer industry, where the pure silicon used is

doped with trace amounts of elements such as gallium and extreme pre-
cautions are taken to ensure that the silicon is pure. Even there, however,
it just means that the level of impurities is below the parts per billion or
trillion levels. The problem can be envisaged by realizing that 1 g of pyrite
contains about 10^{23} (or 1 hundred thousand billion billion) Fe atoms and
twice that number of S atoms. A few million atoms of any element in the
gram of pyrite will really not make much difference. In fact even with
modern analytical methods you will need a million billion atoms of a trace
element to even detect it.

With natural compounds or minerals, we have no control over the
composition, and the crystals contain traces of more or less what was
present in the environment when they were growing. Pyrite crystals,
for example, contain bits of rock, other minerals, and a whole spectrum
of trace elements. This means that when you analyze a pyrite crystal by
standard methods you do not get a ratio of exactly one atom of iron to
two of sulfur. To use Richter's term: the material does not seem to be
stoichiometric.

This idea of nonstoichiometry harks back to the other main protagonist
in the debate on the Law of Definite Proportions. Claude Louis Berthollet
was the doyen of French chemistry in the late 18th century who held the
view that two elements might combine in constantly varying proportions.[37] Although Berthollet was proven generally wrong, there is a class of
materials that show deviations from whole number combinations. These
include the iron sulfides—particularly the pyrrhotites, the magnetic iron
sulfides synthesized by Proust and analyzed by Hatchett. These minerals display a degree of nonstoichiometry whereby the ratios of iron:sulfur
may vary by more than 10%. In the 20th century it became accepted that
pyrite was also nonstoichiometric since classical chemical analyses of natural pyrite crystals gave variable Fe:S ratios. The uncertainties associated
with conventional sulfur analyses together with the impurities intrinsic
in the reagents meant that even synthetic pyrites showed compositional
variations.

The problem was initially resolved by the invention of the petro-
logical reflected-light microscope where opaque materials such as
pyrite are highly polished and their surfaces microscopically viewed.
The reflected-light microscope is an extremely powerful tool, since its
resolution is limited only by the wavelength of the incident light used.
This contrasts with other forms of optical microscopy where diffraction
effects in particular limit the maximum resolution. Since the wavelength

of visible light goes down to about 400 nanometers, this means that it is possible to resolve submicrometer structures (i.e., less than 1,000 nanometers, or a millionth of a meter) in a reflected-light microscope. As a postgraduate student I happily viewed and photographed submicron pyrite structures with reflected-light microscopy and took great pleasure in giving lectures in front of these at international conferences where they were blown up to more than a meter across on the screen behind me.

These techniques showed that apparently pristine crystals of pyrite commonly contain inclusions of other minerals, even down to sizes of one-millionth of a meter. Even then there seemed to be traces of other elements in the pyrite structure that appeared to be included in the pyrite structure and not to be in the form of discrete mineral grains. This problem was resolved by the invention of the analytical electron microscope whereby analyses of areas of the surface less than one-millionth of a meter could be collected.

Gunnar Kullerud and Hatten Yoder of the Carnegie Institute (Washington, D.C.)[38] originally suggested that the composition of pure pyrite is stoichiometric FeS_2. They concluded that reports of deviations from stoichiometry were caused by analytical uncertainties or the presence of traces of other elements in the material. This is the situation today. Pure pyrite is FeS_2, but all natural pyrite contains impurities. No systematic understanding of trace elements in pyrite is available. There have been many studies of pyrite compositions, especially over the past fifty years, but none of these has led to any robust theory as to what controls the trace element compositions of natural pyrites other than they happened to be in the environment when pyrite crystallized and had not been entirely rejected by the mineral since that time. One of the most extensive modern databases of the trace element contents of pyrite has been compiled by the Australian Research Council Centre of Excellence in Ore Deposits at the University of Tasmania over many years. These samples have been screened texturally and analyzed using a microscopic laser beam with the resultant vaporized material sent through a mass spectrometer. This method reduces risks of contamination by impurities in the pyrite. The results suggest that the trace element contents of sedimentary pyrite mainly track the trace element contents of the contemporary seawater and thus might be used as a probe for ancient Earth marine environments. Even so, this compilation has revealed no systematic relationship between trace elements and pyrite.

Notes

1. R. Campbell Thompson. 1936. *Dictionary of Assyrian Chemistry and Geology* (Oxford: Oxford University Press), 266pp; On some Assyrian minerals. 1933. *The Journal of the Royal Asiatic Society of Great Britain and Ireland*, 4:885–895. *The Pennsylvania Sumerian Dictionary*. updated 2006 (Philadelphia: University of Pennsylvania Museum of Archaeology and Anthropology) translates *za* as stone and *bil* as burning.

2. Much of this comes from E.C. Caley and J.F.C. Richards. 1956. *Theophrastus on Stones*. Graduate School Monograph 1 (Columbus: Ohio State University). I could not find a modern English translation of Theophrastus (in the days before the Internet) and started my own in Stockholm in 1975. However, having found the Caley and Richards translation and commentary, happily I did not have to continue the task.

3. Lusitano Amato. 1557. *Pedacii dioscori dae anazarbensis de materia medici, libri V* (Venetiis: Ex Officio Jordani Zilleti), 89pp. In an expansive footnote, Amato quotes Claudius Galen (130–200 CE), the surgeon to the gladiators, as reporting that Archimedes (c. 287 BCE) first described the mineral. Even though this appears to postdate Theophrastus, it suggests that the mineral was well known to Greco-Roman scientists.

4. J. Bostock and H.T. Riley. 1855. *Pliny the Elder, The Natural History* (London: Taylor and Francis). This is a standard translation of the Latin text. Pyrites are mentioned in Book 36 (*The Natural History of Stones*), Chapter 30 ("Millstones, Pyrite, Seven Remedies"):

   > Molarem quidam *pyriten* vocant, quoniam plurimus sit ignis illi, sed est alius spongiosior tantum et alius etiamnum *pyrites* similitudine aeris. in Cypro eum reperiri volunt metallis, quae sint circa Acamanta, unum argenteo colore, alterum aureo. cocuntur varie, ab aliis iterum tertiumque in melle, donec consumatur liquor, ab aliis pruna prius, dein in melle, ac postea lavantur ut aes. usus eorum in medicina excalfacere, siccare, discutere, extenuare et duritias in pus vertere. utuntur et crudis tusisque ad strumas atque furunculos.
   >
   > *Pyritarum* etiamnum unum genus aliqui faciunt plurimum ignis habentis. quos vivos appellamus, ponderosissimi sunt, hi exploratoribus castrorum maxime necessarii. qui clavo vel altero lapide percussi scintillam edunt, quae excepta sulpure aut fungis aridis vel foliis dicto celerius praebet ignem.

5. V. Scheil. 1918. Vocabulaire de pierres et d'objets en pierre. *Revue d'assyriologie et d'archéologie orientale*,15:115–125. R. Campbell Thompson. 1936. *Dictionary of Assyrian Chemistry and Geology*; On some Assyrian minerals. 1933.

6. The complete name was *Marhasi-ki*, where *ki* is the determinant meaning land. Of course the origin of the name *Marhasi* itself should be the next stage in

our quest to find out where the name *marcasite* came from. It was variously transcribed as *Marhashi, Marhasi, Parhasi, Barhasi, Waraḫše,* and *Warakshe.* The existence of Marhasi was established by P. Steinkeller. 1982. The question of Marhasi: A contribution to the historical geography of Iran in the third millennium BC. *Zeitschrift für Assyriologie,* 72: 237–265. According to *The Assyrian Dictionary* (Chicago: University of Chicago, 1956), *marhasu* simply meant "stone" in old Akkadian. This dictionary also gives the meaning of *mar.ha.ši* as carnelian, a red variety of quartz.

7. Dr. Benjamin Hallum of the University of Warwick pointed this out to me. Dr. Hallum is currently working on a translation of the Arabic versions of Zosimos' *On Sulfur* and Hunayn's translation of Dioscorides. As an aside, my Arabic guru Kurda Saied-al-Berezanchi pronounces the *u* in *burites* much more like a short *er* rather than *ur,* so that it sounds closer to *pyrites.* These examples illustrate the problems of interpreting ancient literature: the scribes simply wrote down what the words sounded like to them. The reader then pronounces the written words according to his or her experience, which may not be the same as that of an ancient scribe or, as in my case, the same as a non-Arabic speaker.

8. P. Kraus. 1942–1943. *Jâbir ibn Hayyân: Contribution à l'histoire des idées scientifiques dans l'Islam: I. Le corpus des écrits jâbiriens; II. Jâbir et la science grecque* (Cairo: Institut français d'archéologie orientale). Repr. by Fuat Sezgin. 2002. *Natural Sciences in Islam,* Vol. 66 (Frankfurt: Institute for the History of Arabic-Islamic Science, Johann Wolfgang Goethe University, 200pp), pp. 67–68.

9. *Balinas* is the Arabic name for Apollonius of Tyana, and there are several commentaries on Balinas credited to Jabir. Apollonius of Tyana is another mysterious figure: a wandering Greek philosopher and (some say) miracle worker who rose from the dead, was a contemporary of Jesus of Nazareth, and has figured as a leading figure in many religious cults. He was much admired in early Islam, however, partly because of the power of its talismen, carved figures set up on columns to protect towns. This somewhat heterodox view of Islam may be one reason why the corpus of 10th-century works containing the Balinas hagiographies was attributed to the long-dead Arabic sage, Jabir. Jabir's *Book of Stones* is infamous since it contains the statement (4:12) about the author's reason for writing and publishing his books: "The purpose is to baffle and lead into error everyone except those whom God loves and provides for." Because Jabir's works rarely make sense, the word *gibberish* originally referred to his writings. See E.J. Holmyard. 1928. *The Arabic Works of Jabir ibn Hayyan.* Translated by R. Russel in 1678 (New York: E.P. Dutton).

10. A.Y. Al-Hassan Gabarin. 2002. *An Eighth Century Treatise on Glass. Kitab al-Durra al-Maknuna (The Book of the Hidden Pearl) of Jabir ibn Hayyan (c. 721–c. 815): Part 1. The Manufacture of Coloured Glass* (http://www.

history-science-technology.com/articles/articles%209.html, accessed October 2014).

11. See J. Ruska. 1912. *Das Steinbuch der Aristoteles* (Heidelberg: Carl Winters Universitätsbuchhandlung):

> There are many species of marcasite including gold-, silver- and copper marcasite. If marcasite is calcined and burnt until it becomes a fine flour, it is used in chemistry. Add a little to sulfur in a crucible and it purifies gold. And when it is treated with water and struck with iron it ignites. (my translation)

12. H.M. Said. 1989. *Al-Beruni's Book on Mineralogy: The Book Most Comprehensive in Knowledge on Precious Stones* (Islamabad: Pakistan Hijra Council), 375pp.

13. Note that pyrite is, of course, not black. This may refer to black sulfur, which is bitumen or naptha contaminated with pyrite and black in color and is described by Al-Razi. It is possible that this is equivalent to Theophrastus's *spinos* or Thracian stone.

14. "As to the claims of the alchemists it must be clearly understood that it is not in their power to bring about any true change in species." From J. Linden Stanton, ed. 2003. *The Alchemy Reader: From Hermes Trismegistus to Isaac Newton* (Cambridge, UK: Cambridge University Press). Avicenna's *De minerabilus* was ascribed to Aristotle until 1927, when it was shown that it originated from Avicenna's *Kitabal-shifa* written at Hamaden 1021–1023 CE.

15. Arnold of Saxony listed eighty-one stones in the third part of his book, *The Purposes of Natural Things.* See V. Rose. 1875. Aristoteles De lapidus und Arnoldus Saxo. *Zeitschrift für Deutches Alterthum*, 18:321–455. This appears to have been lifted more or less intact from the *Book of Stones* of Marbode, Bishop of Rennes in 1096. Marbodus Redenensis. 1893. *Liber lapidum seu de gemmis and other writings.* Edited by J.P. Migne. *Patrologie Latinae*, 171:1735–1780. This was the most famous lapidary of the Middle Ages and listed sixty stones. The Franciscan Bartholomew of England wrote an encyclopedia, *On the Properties of Things,* in 1230 CE, which included one book of minerals; Bartholomaeus Anglicus. 1488. *De proprietatibus rerum* (Heidleberg: Lindelbach). Thomas of Cantimpre knew Albert and wrote a similarly titled encyclopedia in 1244 CE; L. Thorndike. 1963. More manuscripts of Thomas of Cantimpre *De naturis rerum. Isis,* 44:269–277. The book on stones is similar to Arnold's.

16. I. Draelants. 2000. *Un encylopédiste méconnu du XIIIIe siècle. Arnold de Saxe. Oeuvres, sources et réception* (PhD diss., Université catholique de Louvain).

17. Marbodus Redenensis. 1893. *Liber lapidum seu de gemmis and other writings.* Edited by J.P. Migne. *Patrologie Latinae*, 171:1735–1780.

18. D. Wyckoff. 1967. *Albertus Magnus: Book of Minerals* (Oxford: Clarendon Press).

19. A modern translation and commentary on this is I. Draelants. 2007. *Le Liber de virtutibus herbarium, lapidum et animalium Liber aggregationis* (Florence: Sismel).

The book was attributed to Albertus Magnus, but it is a later compilation from an unknown author.

20. G. Agricola. 1530. *Bermanus, sive De re metallica* (Basle: In aedibus Frobenianis), 135pp.

21. G. Agricola. 1546. *De natura fossillum.* Translated by M.C. Bandy and J.A. Bandy, 2004 (Mineola, NY: Dover).

22. *A Lexicon of Alchemy or Alchemical Dictionary by Martin Rulandus the Elder (sic).* 1893. Translated by A.E. Waite (London: John M. Watkins). His son Martin Ruland the Younger actually wrote the dictionary and Waite was mistaken in his attribution.

23. And to this day. The present standard model suggests that most energy in the universe is missing. Physicists call this "missing material dark energy." However, some 21st-century physicists are now calling it "quintessence."

24. G. Agricola. 1546. *De ortu et causis subterraneorum.* Translated by H. Hoover and L. Hoover (Basileae: Froben). Hoover and Hoover translate *succus* as *juice* by analogy with the Germanic *saft* as in "fruit juice."

25. R. Boyle. 1664. *Experiments and Considerations Touching Colours* (New York: Johnson Reprint, 1964), 135pp.; and *Short memoirs for the natural experimental history of mineral waters.* 1685. In *Collected Works*, Vol. 4, edited by A. Millar (London, 1744), pp. 237–239.

26. M. Lister. 1682. *De fontibus medicinalibus Angliae* (London: Walter Kettilby). See A.M. Roos. 2004. Martin Lister (1639–1712) and fool's gold. *Ambix*, 51:23–41.

27. Lister had implicated pyrite in the origin of thunder, lightning, and earthquakes because of the "inflammable breath of the pyrites, which is a substantial sulphur, and takes fire in itself." Roos. 2004. Martin Lister (1639–1712) and fool's gold. *Ambix*, 51:23–41.

28. A.M. Roos. 2004. Martin Lister (1639–1712) and fool's gold. *Ambix*, 51:23–41.

29. J.K. Wallerius. 1747. *Mineralogia eller mineral-riket, indelt och beskrivit af J.G.W.* (Stockholm: Lars Salvius). This is often regarded as the first modern mineralogy textbook; it ran into several editions and was translated into German and French.

30. J.B. Richter. 1892–1894. *Anfangsgrunde der Stöchyometrie oder Messkunst chymischer Elemente* (Breslau: Hirschberg).

31. L. Proust. 1794. Les sulfures natifs et artificiels du fer. *Journal de Physique, de Chimie et d'Histoire Naturelle*, 54:89–96. One of the curious things about this French publication is that its date is given as "Pluviose an 10." This is the date according to the Revolutionary calendar, which renamed the months and years from 1792 = year 1. *Pluviose* was more or less equivalent to February. The problem here is that 1794 should be year 3.

32. C. Hatchett. 1804. An analysis of the magnetical pyrites; with remarks on some other sulphurets of iron. *Philosophical Transactions of the Royal Society of London*, 94:315–345.

33. J. Dalton. 1808. *New Theory of Chemical Philosophy*, Vol. 1 (Manchester, UK: R. Bickerstaff), 560pp.

34. J.J. Berzelius. 1813–1814. Essay on the cause of chemical proportions, and on some circumstances relating to them: together with a short and easy method of expressing them. *Annals of Philosophy*, 2:443–454; 3:51–62, 93–106, 244–255, 353–364. This paper was printed in several sections over two years.

35. J.J. Berzelius. 1819. *Essai sur la théorie des proportions chimiques et sur l'influence chimique de l' électricité* (Paris: Méquignon-Marvis), 337pp.

36. J.J. Berzelius. 1820. Mineranalysen. *Schweiggers Journal*, 27:67–76.

37. C.J. Berthollet. 1803. *Essai de statique chimique*, Vol. 1 (Paris: Firmin Didot), 543pp.

38. G. Kullerud and H.S. Yoder. 1959. Pyrite stability relations in the Fe-S system. *Economic Geology*, 54:533–572.

4

Crystals and Atoms

Introduction

According to one magic crystal website, pyrite is a highly protective stone blocking and shielding you from negative energy. This may originate from Pietro Maria Canepario, who in 1619[1] cited Avicenna as stating that "if pyrite is worn on an infant's neck, it defends him from all fear."

Other New Age sources maintain that pyrite can be beneficial when planning large business concepts because placing a piece on the desk energizes the area around it. Pyrite also reduces fatigue and is good for students because it is thought to improve memory and recall and to stimulate the flow of ideas. So you are certainly reading the right book . . .

The magical properties of pyrite stem at least partly from the occurrence of pyritized ammonites (Figure 4.1) in ancient Egypt. Ammonites are fossils of coiled mollusks that became extinct at the same time as the dinosaurs at the end of the Mesozoic Era, about 60 million years ago. Ammonites got their name because they resemble coiled ram's horns and the Egyptian god Amun (or Amon, Ammon, etc.) usually wore ram's horns. The person responsible for this flight of fancy was Pliny the Elder, who called these fossils *ammonis cornua* or horns of Ammon. The golden pyritized ammonites were prized as lucky charms and worn as amulets in ancient Egypt. They are common today and may be readily collected from the beach at Charmouth in southern England, particularly after a storm has caused more fresh rock from the cliffs to tumble down onto the beach.

The bright golden crystals of pyrite have fascinated humankind through the ages. The crystals display a variety of distinct shapes that make them extremely attractive. Indeed, pyrite may display the greatest variety of crystal forms of any common mineral. The great American mineralogist

FIGURE 4.1. Pyritized ammonite from Charmouth, England (see color plate). Photo M. Keating.

James Dwight Dana described eighty-five different forms, and the founder of geochemistry, Victor Moritz Goldschmidt, drew line drawings of almost 700 different pyrite crystals. In this chapter I show how the explanation of this extraordinary diversity of pyrite crystal shapes (or *habits,* formally) has helped reveal the nature of the material universe.

As discussed in Chapter 3, crystals are made up of atoms and molecules, but it was not until the last decade of the 20th century that technological developments permitted atoms to be actually seen. The problem is twofold. First, atoms are very small. The crude approximations to atomic sizes have been based on the billiard-ball approach to the rendering of the shapes of atoms. If we assume that atoms are shaped like billiard balls, then their sizes can be estimated by considerations of crystal structures. On this basis, for example, the iron atom in pyrite is just 75 picometers in diameter; that is, 75,000,000,000 Fe atoms would reach 1 millimeter. The sulfur atom is about twice as big as the iron atom.

When we start looking at atoms we enter a strange *quantum world* where cause and effect no longer apply. *Quantum,* the Latin word for *amount*, describes the tiny particles that make up matter and refers to the smallest discrete quantities of energy that can exist. This quantum world is the real world and our everyday experience of order and of cause and effect is just apparent. Our actual existence is really a scary, Alice-Through-the-Looking-Glass reality where chance rules. In this world, particles can move instantaneously, can be in two places at the same time, and only appear to behave themselves normally when we look at them, rather like naughty children in a classroom. The quantum world is more than just a philosophical construct. Its principles are what run computers and mobile phones, televisions and lasers, and all electronic devices. Before the advent of quantum mechanics in the early years of the 20th century, the world was ruled by straightforward Newtonian physics, modified by Einstein. This approach is a good approximation of the behavior of materials like pyrite and is called the classical approach. You will be relieved to know that we use the classical approach in this book. Pity, though: the quantum world, like the Cheshire Cat, is much more fun, although ultimately threatening.

Atoms were originally described in the prequantum, classical world by examining crystals and in this pyrite played a key role. In this chapter I explore the legion of crystal forms displayed by pyrite and explain why pyrite shows this extraordinary variety of forms. In doing this, we see how the study of pyrite crystals has provided a foundation for ideas as diverse as mathematical topology, the origins of symmetry, as well as the atomic nature of matter.

Pyrite Habits

The nearest geology has to a saint is Niels Stensen, or Steno, who in his spare time in Florence as a novice priest devised the primary Law of Crystallography. Steno has been beatified, which is the first step to sainthood in the Catholic Church, and his feast day is December 5. He is the only geologic saint, although his sainthood was not due to his crystallographic work but rather to his contribution to the Church as Archbishop of Munich.

In 1669 Steno published *De solido intra solidum naturaliter contento dissertationis prodomus*, a book about natural occurrences of stones within stones, a volume with four parts. In the first part, Steno describes marine objects, including fossil sharks' teeth embedded in rocks and found at a distance from the sea. He confirmed that fossils were the remains of living creatures. This idea of fossils in rock drove the book's central concern of how solids are formed within solids. Part 3 of the book addresses solids occurring within solids, and the volume concludes with Part 4, a discussion of the geology of Tuscany in terms of the effects of Noah's Flood. He defines the basic principles of stratigraphy, which included the concept that the fossil record represented an ordered succession of different creatures. This work was fundamental to Darwin's later ideas on evolution.

However, it is the second part of *De solido intra solidum naturaliter contento dissertationis prodomus* that concerns us here. In this Steno states that the angles between corresponding faces on crystals are the same for all specimens of the same mineral. This is an extraordinary idea. It explains how crystals of a mineral, like pyrite, can display hundreds of shapes but the angles between the crystal faces that confine these shapes are all the same. It suggests that there is some underlying control in the molecular structure. This is the basis of crystallography and is known as Steno's Law.

As I show next, simply inspecting pyrite crystals can provide insights into the nature of their internal structure on the atomic scale. This is conceptually astonishing: it means that you (and I mean you, the reader) can deduce atomic structures without actually being able to see the atoms.

Johann Friedrich Henckel[2] recognized in 1725 that pyrite crystals occurred in a variety of forms and tried to classify them (Figure 4.2). He listed tetrahedra, pentahedra, cubes, rhombs, octahedra, decahedra, dodecahedra, and prisms. If he had been able to measure the angles between the crystal faces accurately, he would have found that all these forms were combinations of cube, octahedral, and irregular dodecahedral faces: they all obeyed Steno's Law. Henckel understood that there was something about these crystals that was highly significant in terms of the basic constituents of these materials. His classification is similar to that of the contemporary botanist, sorting plants out according to their shapes without understanding the genetic code. It constituted a step change in humankind's approach to mineralogy from the simple listings of varieties of minerals and rocks according to their source, color, or magical properties, which had culminated in 1612 with Rulandt's listing of seventy-one species or varieties of pyrite described in Chapter 3.

Steno's rules on the constancy of crystal angles were conceptualized by the Abbé Haüy in 1801.[3] He pointed out that crystals of the same species assume very diverse forms but all had the same basic structural element. He noted with respect to pyrite: "The sulfide of iron or pyrite produces

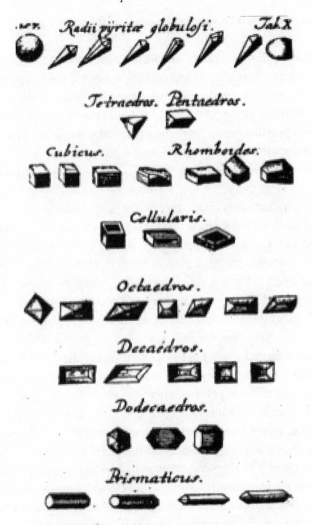

FIGURE 4.2. Henckel's classification of pyrite crystal forms in 1725.

now cubes, now regular octahedra, here dodecahedra with pentagonal faces (pyritohedra), there icosohedra with triangular faces (pyritohedra and octahedra)."

He went on to demonstrate that crystals were made up of a virtually infinite number of small regular molecular units, which we now call *unit cells*. The unit cell of a crystal is the smallest divisible unit of a crystal that possesses the symmetry and chemical properties of the bulk mineral. *Symmetry* in this context refers to the ability of crystals to appear identical when they are rotated or reflected in a mirror. As you can see, the actual crystal shape may not be very useful since with pyrite, for example, there may be hundreds of different shapes.

By contrast, the number of shapes of unit cells is extremely limited since, in order to be repeated throughout the crystal, they need to be regular so that they can fit together. The packing of these unit cells cannot leave any gaps or spaces as the crystal could not hold together in this case. There are only five possible forms for regular polyhedra (tetrahedron, cube, octahedron, dodecahedron, and icosohedron with four, six, eight, twelve, and twenty faces, respectively; Figure 4.3). These are known as the Platonic solids, since Plato theorized that the classical elements were constructed from these regular solids. The ancient Greeks studied these solid forms in detail, and various authors have been credited with their discovery, including Pythagoras. However, it appears that Plato's contemporary, the mathematician Theaetetus, provided the first proof that there are no other convex regular polyhedra. This was subsequently confirmed by Euclid, or perhaps Euclid plagiarized Theaetetus's work in Book XIII of *The Elements*.[4]

The idea that there are only five possible regular polyhedral solids has fascinated scientists over the millennia. Before he discovered that planetary orbits were ellipses, Kepler had originally argued that distances between the five inner planets were arranged according to the relative sizes of spheres enclosing each of the five Platonic solids in order—and all of them were enclosed within the sphere described by the orbit of Saturn, the only other planet known during Kepler's time. You can see how a relationship between astrology, magic, and crystals developed.

Haüy showed how the various shapes or habits of crystals could arise from stacking of countless microscopic blocks in the form of regular polyhedra. The microscopic blocks are so small that they produce a macroscopically smooth crystal surface so that, for example, a cubic unit cell could be stacked to produce octahedral and pyritohedral faces as well as cubic ones (Figure 4.4). Haüy's insight explained how pyrite could display

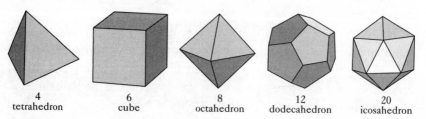

| 4 | 6 | 8 | 12 | 20 |
| tetrahedron | cube | octahedron | dodecahedron | icosahedron |

FIGURE 4.3. The five Platonic solids with the numbers of sides indicated.

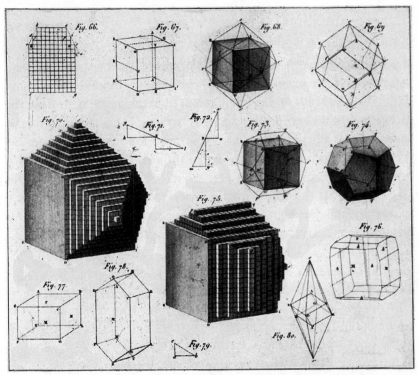

FIGURE 4.4. Haüy's model of how cubic building-block elements of crystals accounts for their regular and consistent shapes. From R.J. Haüy. 1801. *Traite de Mineralogie* (Paris: Louis).

many hundreds of crystal forms and yet still obey Steno's Law about the consistency of interfacial angles.

Topology is an important branch of mathematics dealing with solid shapes and their deformations that has important technological applications today. It developed in the last decades of the 19th century, and crystallography helped to provide the grounding for the science. Haüy's block model of crystal structures provided the basis. We have seen that the microscopic blocks (or unit cells) have simple geometries and that therefore it is reasonable to ask how many combinations of these are possible. These combinations should describe all crystals. In 1891 the Russian crystallographer E.S. Fedorov showed that all crystals could be described by just 230 types of combinations.[5] Fedorov also invented the goniometer, an optical instrument for precise measurements of the angles between crystal faces. Using the goniometer, it was possible for the first time to confirm Steno's original law of the constancy of the angles between crystal faces and to demonstrate that the pyritohedron was not a true dodecahedron.

Pyrite and the Foundations of X-Ray Crystallography

If you look at the surface of a CD or DVD disk at an angle you will see a shimmering spectrum of colors on the surface of the disk as bands of luminous greens and blues seem to radiate out from the center of the disk. The grooves on the disk are diffracting the light that is being reflected from its silver surface. Diffraction occurs when a wave encounters an obstacle. The everyday analogy is that of waves on the surface of a pond or the sea that bend around a rock (Figure 4.5). As the wave hits an object, new waves are produced at all points along the wave front. These waves propagate spherically, and thus light can appear to bend as it passes an object. This bending effect is seen in the shadowy image that appears at the edge of a

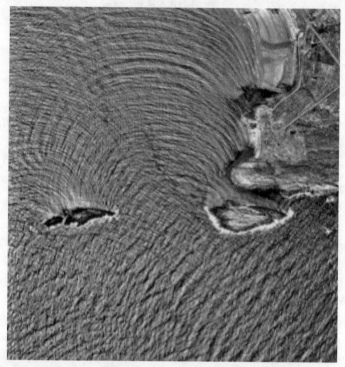

FIGURE 4.5. Diffraction of waves. The waves coming from the bottom of the image pass between two rocks where they are bent or diffracted. The rocks are acting like a diffraction grating. If you look closely you can see apparent radiating lines in the area above the rocks where the diffracted waves are added and subtracted to make linear ridges and troughs. Photo courtesy of the Norwegian Mapping Authority: Fjellanger Widerøe AS.

sharp object. Of course if there is a narrow slit, light will appear to bend around both edges of the split as shown in the split formed by the rocks in the ocean in Figure 4.5. And if the width of the slit approaches the wavelength of the light, the light waves emitted from the split edges will either be in phase or out of phase: if the diffracted waves are in phase (i.e., their peaks and troughs are coincident) then the resultant intensity is increased; if the diffracted waves are out of phase, then the peaks are canceled out by the troughs and no light is seen.[6] Careful examination of the sea to the north of the rocks in Figure 4.5 shows faint radiating lines formed by these troughs and peaks.

In the case of light, the troughs and ridges of the ocean in Figure 4.5 are represented by a series of bands of light. These depend on the wavelength of the incident beam and the density of the slits in the object. The diffraction effect is seen on the fine grooves of a CD disk but not on a grill, for example. In a typical diffraction grating the number of slits ranges from a few tens to a few thousand per millimeter. Note that since there is a relationship between the wavelength of light and the slit width, each wavelength of the incident beam is sent in a slightly different direction. This can produce a spectrum of colors from white light illumination, visually similar to the operation of a glass prism; this is the shimmering, multicolored effect on the CD surface. The upshot of all this is that by measuring the angle of the emitted light from a diffraction grating and its wavelength, we can calculate the size and number of the slits in the grating that produced the spectrum.

In 1912 Max Laue reported that X-rays were diffracted by crystals.[7] These mysterious rays had been discovered by Wilhelm Conrad Röntgen in 1895 and had the ability to penetrate solids. Laue's great insight was to realize that since X-rays have wavelengths similar to that of the distances between atoms in crystals, the atoms would act rather like the islands in Figure 4.5 and diffract the X-ray waves. A crystal should act like a three-dimensional CD, producing bands of more and less intense X-rays. As with the CD and other diffraction gratings, the distances between the X-ray bands and their intensities depend on the distances between the atoms in the crystal. X-rays did not provide just a simple ghost-like view of the object as in medical X-rays but exited in a pattern determined by the atomic structure (Figure 4.6).

The technique was seized upon by W.H. Bragg and W.L. Bragg. The Braggs realized that the angles and wavelength of the X-rays diffracted by a crystal would be functions of the positions of the planes of atoms in

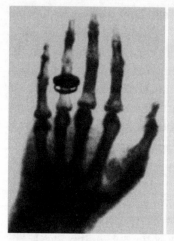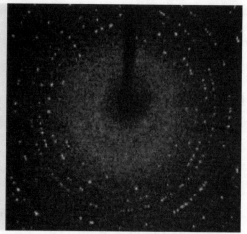

FIGURE 4.6. Difference between medical X-rays, which display differences in X-ray density of the object, and X-rays of pyrite crystals, where the X-rays are diffracted by the crystal lattice structure. The image of Mrs. Röntgen's hand was the first X-ray picture of the human body ever taken, just one week after Röntgen had accidently discovered X-rays (image courtesy of NASA). The image of the X-ray diffraction pattern of the pyrite crystals shows a series of white spots resulting from diffraction of the X-rays by the pyrite crystal lattice. Measurements of the angles and relative distances of the spots reveal the atomic structure of the crystal.

the crystal. Since there are several such planes in any crystal, this would enable the atomic structure of the crystal to be computed. They could see and measure Haüy's unit cells. W.H. Bragg, the father, was the Cavendish Professor of Physics at the University of Leeds. He was interested in the nature of X-rays and used the crystal diffraction phenomenon to test whether X-rays behaved as a series of particles. W.L. Bragg, the younger, was a graduate student at the Cavendish Laboratory. In order to distinguish himself from his father he took to using the name "Lawrence." He was more interested in the application of X-rays to study the structure of materials. In 1913, the duo published a paper describing the application of the X-ray-generating ionization spectrometer developed by the elder Bragg to crystals. This paper first published Bragg's Law, which is the basis for interpreting the atomic structure of materials with X-ray and electron beam methods.

Pyrite was one of the first crystalline materials investigated by the Braggs. They used pyrite to demonstrate that X-rays behaved in the same manner as light and not, as the elder Bragg had previously supposed, as a series of particles. The structure of pyrite, however, proved more difficult to unravel. The younger Bragg published a Nobel Prize-winning paper in

1913 on the atomic structure of common salt, sphalerite, fluorspar, and calcite in which the difficulties with the pyrite structure interpretation were mentioned. Finally, in 1914, Lawrence Bragg succeeded in solving the pyrite structure and confirmed Fedorov's original theoretical results.[8] In particular, Bragg's results demonstrated that the symmetry centers of Fedorov's space groups are occupied by separate atoms. This remarkable confirmation of a theoretical mathematical model by subsequent physical analyses is one of the great, though generally unappreciated, triumphs of inductive science. The distribution of the Fe and S atoms in pyrite was particularly important here, since the space group Fedorov assigned to pyrite were considered an imaginary system, in the same way as the square root of −1 is an imaginary number. In fact, it turned out that the arrangements of atoms in the pyrite unit cell are exactly as predicted by the imaginary Fedorov systems.

Pyrite helped support the foundations of X-ray crystallography because it showed how the method could be used to determine the structure of a substance with a more complex structure than a salt cube. This ultimately led to the determination of the structure of DNA in 1953 by Crick and Watson based on Rosalind Franklin's X-ray crystallographic analyses.

Pyrite Structure: Counting Atoms

The pyrite structure, as originally worked out by Lawrence Bragg, is shown in Figure 4.7. The figure illustrates the pyrite unit cell, the basic building block of the pyrite crystal in terms of Haüy's model. The length of the sides of the pyrite unit cell is 0.5417 nanometers. This means that, in a 1-cm pyrite cube, there are almost 10^{23} (1 followed by 23 zeros) identical unit cells and that the lattice can be approximated as an infinite arrangement of unit cells.

These diagrams stretch the actual molecular structure: the atoms are rendered as small balls, with correct relative sizes, but the distances between them are exaggerated so that you can see the arrangement of the atoms more clearly. A more realistic model would pack the atoms closer together so that the electron orbitals of the sulfur and iron overlap, creating the chemical bonds that hold the material together.

Figure 4.7 shows that pyrite is basically a cubic structure. The corners of the cube are occupied by the smaller iron atoms, which are also situated at the centers of the faces of the cube. This is called a

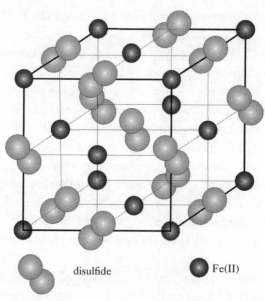

disulfide   ● Fe(II)

FIGURE 4.7. The pyrite unit cell.

face-centered cubic arrangement. This is one of the densest packing arrangement possible and makes pyrite fairly heavy, with a density almost five times greater than water and twice the average rock
density. However, the sulfur atoms are paired and the centers of the
pairs occupy the midpoints of the cube edges and the cube center. Note
also that the sulfur pairs are oriented in opposite directions in alternate planes in the cube. This means that, although pyrite is basically
cubic, its symmetry is lower than the symmetry of an ideal cube with
single atoms in these positions. If you were to rotate a cube with just
the iron atoms in place, the identical arrangements occur four times.
Adding the alternately angled sulfur pairs, or dumbbells, produces
a lower symmetry: rotation of the pyrite unit cell does not give four
identical arrangements. Rather, as we see later, the same arrangement
only occurs twice on rotation of a pyrite unit cell. We begin to see why
Fedorov's original calculated classification suggested that the pyrite
unit cell was only partly symmetric.

Because the unit cells of crystals are adjacent, most of the atoms shown
in Figure 4.7 are shared with adjacent unit cells. There are actually just
four FeS_2 molecules in each pyrite unit cell.[9] Using this information and
the data on the density of pyrite, we can count the number of Fe and S
atoms in a crystal of pyrite. A nice 1-cm pyrite cube contains 25×10^{19} (i.e.,

250,000,000,000,000,000,000,000 or 250 quintillian) Fe atoms and 50 ×
10^{19} (or 50 quintillian) S atoms.

The invention of the scanning tunneling microscope (STM) in 1981
earned Gerd Binning and Heinrich Rohrer the Nobel Prize for Physics
in 1986. The STM uses the weird world of quantum physics to see and
image atoms. The electrons around the iron and sulfur atoms that produce the electronic signals in pyrite are recorded by the STM, and, in
the strange world of quanta, they have equal probabilities of being in the
probe and in the pyrite atoms at the same time. The resulting analysis of
the electronic signals in the received from the probe provides an image of
the atoms on the surface. The STM image of pyrite (Figure 4.8) confirms
the close-packed, face-centered cubic arrangement of the iron atoms in
the structure as computed from the X-ray diffraction pattern some eighty
years earlier and predicted from symmetry calculations by E.S. Fedorov
100 years previously.

FIGURE 4.8. Scanning tunneling microscope image of pyrite showing the Fe
atoms in a face-centered cubic array. The atoms are around 100 picometers (i.e.,
one-billionth of a millimeter) in size. The blurry image is real: atoms are not hard
billiard balls, and you are seeing merely the densest parts of the cloud of electrons
that surrounds each atom. Photo Kevin M. Rosso.

Cubes and Atoms

Cubes are the most common pyrite crystal form. The cubes shown in Figure 4.9 are from the Ampliación a Victoria deposit near Navajún in Rioja, northern Spain. The ore in this mine is pyrite and the pyrite cubes are mined for mineral collectors. Pyrite has been known and mined in this region for centuries, but the Navajún deposit was discovered in 1965 and is the source of most of the splendid pyrite cubes in the world today. The pyrite is about 130 million years old, and the crystals grew in soft, marly rocks when the marls were buried deep in the Earth's crust and heated to around 350°C.[10] The cubic crystals have grown naturally: they have not been sawn or cut or polished. The largest reach 19 cm in side length.

Steno's Law states that merely the angles between the cubic faces on a pyrite crystal are always at right angles, so that not that all pyrite crystals with cubic faces need actually to be cubes. Some faces may grow faster than others, leading to prismatic forms with identical cubic ends and flat

FIGURE 4.9. Pyrite cubes from Navajún, Spain. These 5-cm crystals are in their natural state: they have not been cut or polished (see color plate).

FIGURE 4.10. Striated pyrite cubes from Huanzala, Mexico. Photo James Murowchick.

faces. In the most extreme case pyrite wires have been observed where the faces at the top and bottom of the cubes have grown so fast that elongated, wire-like crystals have been produced.[11]

The Navajún pyrite cubes are rather special since they display bright, smooth faces. Pyrite cubes are generally striated (Figure 4.10), and the origin of these striations has been the subject of much discussion. Examination of the striations in Figure 4.11 helps resolve the problem. If you take a simple cube and rotate it, the same geometry will appear four times—this is the basic principle of dice games. All four situations are identical, and there is an equal chance of any of the four faces turning up. We can say that the pyrite cube shows fourfold symmetry. But look at the striated cube. The striations on each face are perpendicular to those on the adjacent face. So now when you rotate the striated cube, the same geometry comes up only twice in a rotation. That is, the fourfold symmetry of the smooth cube has been reduced to only twofold symmetry in the striated cube. This reduction in symmetry must mean that the striated cube does not consist of only cubic faces but must involve some faces of lower symmetry.

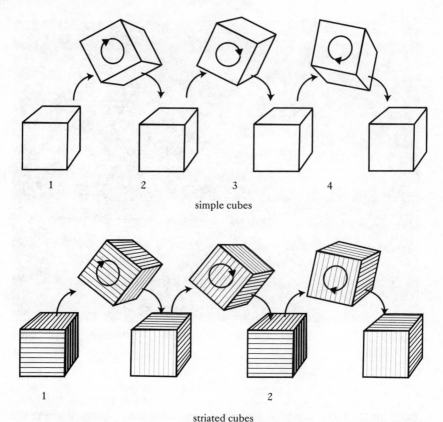

simple cubes

striated cubes

FIGURE 4.11. Comparison of rotational symmetries of simple and striated cubes. Simple cubes can be rotated with four identical results; rotation of striated cubes returns to the same situation only twice.

This means that the unit cells of pyrite shown in the Abbé Haüy's drawings in Figure 4.4 cannot simply have iron and sulfur atoms at their corners or in their centers; otherwise when you rotated them they would come up with the same arrangement four times. So what you must have is a pair of sulfur atoms at the corners of the cube, since pyrite is FeS_2. And these pairs of sulfur atoms must be arranged in such a way that (a) the centers of the pairs are exactly on the corners of the cube, in order to retain the overall cubic symmetry, and (b) they must be arranged in such a way that they come to identical arrangements twice every rotation. So now have a look at Figure 4.7 again: you have actually proved the atomic structure of pyrite without resorting to any expensive X-ray diffraction systems or atomic microscopes.[12] This demonstrates the power of Steno's original insight: we can gain information on the atomic makeup of crystals by simple, careful inspection. Your analysis of these pyrite cubes demonstrates that crystals must be made up of myriads of tiny identical cubes, as the Abbé Haüy reasoned. The striated pyrite cubes show that these tiny identical cubes must be made up of atoms; otherwise all these cubes would be the same. This analysis of these pyrite cubes also demonstrates something else about science: proof of the basic concepts, such as that matter is made up of atoms, is readily accessible to everyone.

Dodecahedra Exist, Almost

The next most common crystal habit of pyrite is the pentagonal dodecahedron (Figure 4.12), which, because of its peculiar relationship to pyrite, is known as the *pyritohedron*. Pentagonal dodecahedra have twelve faces, each of which is pentagonal.

The dodecahedron was known long before Plato's time. Dodecahedral pyrite crystals are common in Italy, and artificial dodecahedra were made there before 500 BCE. The wonderful Miss Dorothy Nairn Marshall of Kames Garden Cottage, Port Bannantyne, Isle of Bute, Scotland, described 387 carved stone balls found in Scotland dating from the Late Neolithic to Early Bronze Age, around 1,000 years before Plato. Five of these were

FIGURE 4.12. Pyritohedra. Although these are the nearest forms to the dodecahedron, inspection of the large crystal on the right, for example, shows that the faces are not the same size nor are they regular pentagons (see color plate).

claimed by some authors to represent the five Platonic solids. Marshall reported that the 387 stone balls had between 3 and 135 knobs on them, so it is likely that these symmetric objects were carved randomly and that some would approximate, by chance, the geometry of the Platonic solids.[13] Even so, it seems probable that, in view of the abundance of pyritohedra, the ancients were aware of and even treasured the dodecahedral form.

However, closer inspection of the pyritohedron in Figure 4.12 shows that the pentagonal faces in the pyritohedron are not regular, since five-fold symmetry is not possible for convex regular polyhedra. To understand why this is and why this affects crystals, we need to return to the Abbé Haüy and the smallest building blocks of crystals, the unit cells. One of the consequences of the Abbé's insights is that since these building blocks are so small relative to the crystal size, crystals are essentially infinitely continuous assemblages. To achieve this the unit cells need to be packed together with no spaces between them. If we look at the simple planar regular polygons, the triangle, square, pentagon, and hexagon, we can see that they can all be packed together, leaving no spaces (Figure 4.13), except for the pentagons, where there is a gap.[14] Basically, this means that normal crystals cannot have fivefold symmetry since this would leave spaces in the structure. The other Platonic solids in Figure 4.3, the dodecahedron and icosahedron, also have fivefold symmetry: the faces of

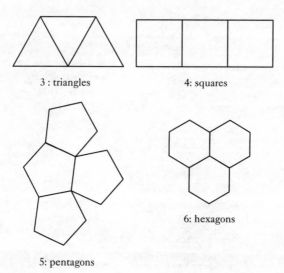

3 : triangles

4: squares

6: hexagons

5: pentagons

FIGURE 4.13. Packing (or tessalation) of regular polygons. Triangles, squares, and hexagons pack with no spaces left; packing of regular pentagons, by contrast, leaves spaces.

the dodecahedron are pentagons and the groups of five triangular faces of the icosahedron, which meet at the vertices of the icosahedron, form a pentagon. The dodecahedron and icosahedron are also, therefore, forbidden symmetries.

So what about the pyritohedron? It cannot be a conventional regular crystalline form since it has forbidden fivefold symmetry. The pyritohedron is believed to have provided Plato with the idea of the dodecahedron, but if he had been able to measure the interfacial angles of the pyritohedron accurately he would have found that they are not all 102°. That is, the pyritohedron is not a regular dodecahedron. However, precise measurements of its interfacial angles would have to wait 2,500 years until Fedorov's invention of the goniometer.

Octahedra and How Pyrite Crystals Grow

Pyrite octahedra are the least common of the simple habits displayed by natural pyrite crystals. In fact, natural octahedra are quite rare, although we can synthesize small octahedra readily in the laboratory. As can be seen in Figure 4.14, the octahedra are not, in fact, all perfect: the tips are flattened off. These flattened faces are cubic faces, and octahedra modified by cubes in this way are a common pyrite habit.

The reason for this was first suggested by the great Japanese mineralogist Ichiro Sunagawa in 1957, and I further developed this in 2012.[15] In order for a crystal to grow, the concentrations of the dissolved constituents must exceed the solubility product of the mineral. The solubility product is the product of the concentrations of all the components of the solid material in solution; in the case of pyrite, for example, it is the product of the concentrations of dissolved iron and sulfur. For pyrite to nucleate from solution, the concentration of dissolved iron and sulfur must at least equal the solubility product. The amount by which the concentration in solution exceeds the solubility product is called the *supersaturation,* and this can be regarded as a measure of the chemical energy available to create the solid pyrite. If the concentration of these constituents falls below the solubility product, the mineral will dissolve.

Pyrite is a very insoluble mineral, so its solubility product is very low—indeed the solubility product is so low that the concentrations of iron and sulfur in solution are virtually immeasurable. Nucleation is the key here: it refers to the first stage of the formation of crystals where atoms and molecules initially coalesce to produce the first molecular

FIGURE 4.14. Pyrite octahedra from the Huazala Mine, Huallanca District, Huanuco Department, Peru (see color plate). Photo Carlos Millan.

embryos of the mineral, which can then grow by crystal growth. However, as with many poorly soluble minerals, pyrite will not nucleate from solution unless the solution is heavily supersaturated with respect to iron and sulfur. The solution must be supersaturated with iron and sulfur by more than 100 billion times the solubility product at 25°C for pyrite to spontaneously nucleate. That is, there must be 100 billion times more iron and sulfur in solution than the equilibrium concentration. This brings the concentrations of iron and sulfur in pyrite-nucleating solutions to measurable concentrations.

Having nucleated, the degree of supersaturation then determines how fast the crystals grow. The energetics of the growth of the pyrite cube faces, octahedral faces, and pyritohedral faces are different. The pyritohedral face requires the greatest amount of energy and thus tends to be preferred at the highest supersaturations. The octahedral face is the next highest, and the most stable cubic face the least. So in a situation where the supply of nutrients is limited (or slower than the rate of crystal growth), crystal growth depletes the concentration of the dissolved components and the crystal faces change with time. In the case of the octahedral crystals, these will grow until the nutrients in solution are used up and then the cubic faces will take over. So most octahedra are capped by cube faces.

Pyrite Raspberries

One of the most common forms of pyrite in nature is as small, globular aggregates of pyrite crystals called *framboids*, since they look like tiny raspberries (Figure 4.15).[16] Pyrite framboids are mostly invisible to the naked eye with diameters usually around 0.01 mm. However, occasional giant ones reach 0.1 mm in diameter, and these can be seen with a hand lens. Framboids are found in rocks, especially sediments, of all ages. The oldest reported pyrite framboids may be from 2.9-billion-year-old sediments from South Africa. They are therefore extremely stable configurations and can last over eons of geologic time.

The abundance of pyrite framboids is quite extraordinary. A guesstimate of the total number of framboids in the world suggests that there are around 10^{30} (1,000,000,000,000,000,000,000,000,000,000), or 10 billion times the number of sand grains in the world, or about 1,000,000 times the number of stars in the universe.[17]

The small size of framboids prevented early workers from discerning their internal organization. They were first described in the 19th century as microspherules and various French workers recorded them in a number of

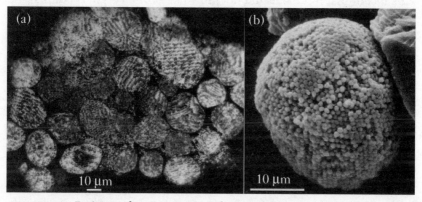

FIGURE 4.15. Pyrite raspberries. (a) Typical view of a section through a group of pyrite framboids in reflected light microscopy. Note the various ordering patterns of the individual pyrite crystals. (b) Detail of a pyrite framboid in showing typical subspherical form and partial ordering of 0.001-mm pyrite crystals (scale 1μm = 0.001 mm). From D. Rickard. 2012. *Sulfidic Sediments and Sedimentary Rocks* (Amsterdam: Elsevier), 801pp. Reprinted by permission.

esoteric environments, including a Roman pavement and, much to their obvious delight, in the timbers of Queen Victoria's royal yacht, *Osborne*. In the early 20th century, improved microscopy showed that these spherules consisted of aggregates of pyrite crystals less than 0.001 mm in size. So each framboid may contain over 1 million tiny crystals of pyrite.

The extraordinary feature of framboids is that each of these 1 million microscopic pyrite crystals has a similar shape and size. Not only that, but they are often beautifully organized and arranged in the framboid. This led to many ideas of how framboids formed, including the theory that they were fossilized microorganisms, as discussed in Chapter 9.

They are eye-catching features under the microscope since pyrite, when polished, reflects light very effectively and the framboids gleam against the dark background of the host rock. This is one of the reasons that so many research students and their supervisors have spent so much time examining framboids and speculating on what they actually represent and how they are formed.

In fact, it was only in the 21st century that detailed studies by my group revealed that framboids were not truly spherical but had flattened faces and that the internal organization of the pyrite microcrystals in framboids was icosahedral. As discussed earlier in this chapter, icosahedral symmetry is forbidden in crystallography and detailed X-ray diffraction studies showed that, indeed, framboids were not single crystals and the individual pyrite microcrystals were not all crystallographically orientated. This meant that framboids did not grow like normal crystals but that the individual pyrite microcrystals aggregated together under the influence of their surface electrical charges. Since these crystals are so small, with 50 million of them usually needed to make up 1 gram of pyrite, these tiny surface electrical forces are sufficient to stick the crystals together. We can see this effect in clumps of pyrite nanocrystals that have failed to come together as individual spheres. We call this *pyrite dust* (Figure 4.16), and it is readily synthesized in the laboratory where access to the site of pyrite nucleation is not restricted by sediments. The pyrite dust shows patches of pyrite nanocrystals that have aggregated into arranged and ordered zones and other areas where no discernible organization has occurred. In fact, entirely regular arrangements of pyrite microcrystals are relatively unusual since it only takes one to be imperfect or an impurity to be caught up in the nucleation process for irregularity to ensue.

The formation of the framboid texture requires that over 1 million pyrite crystals be formed in the same space at the same time and that they

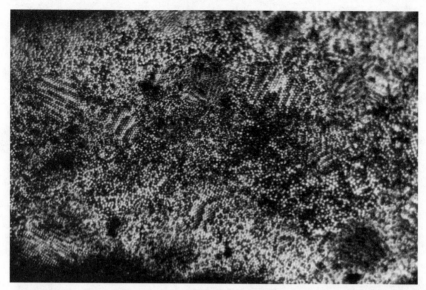

FIGURE 4.16. Pyrite dust from a modern sediment. This is a section through a mass of pyrite nanocrystals, each less than 0.001 mm in size, which have failed to aggregate into spheres. However, some patterns can be discerned that represent zones where the nanocrystals have aggregated regularly. From D. Rickard. 2012. *Sulfidic Sediments and Sedimentary Rocks* (Amsterdam: Elsevier), 801pp. Reprinted by permission.

do not grow much larger than about 0.001 mm. That is, when they reach 0.001 mm in size there is insufficient dissolved iron and sulfur left in the solution for the crystals to grow any further.

As discussed earlier, the key characteristic of pyrite formation from aqueous solutions is the extremely high degree of supersaturation of dissolved iron and sulfur required for pyrite crystals to nucleate. This is a double-edged sword. It means that dissolved iron and sulfur contents in solutions can reach extraordinary high and unstable levels before they hit the tipping-point for pyrite to precipitate. This tipping-point is called a *nucleation burst,* and it describes how the millions of tiny pyrite crystals are suddenly formed in solution, like the starburst of a firework. So framboids result from nucleation starbursts, which seems an apt description of these tiny bright golden spheres in a dark host rock. We can imagine looking into a sediment in which pyrite is forming and watching the pinpoint flashes of gold as pyrite nucleation bursts occur in different places at different times.

Of course once the nucleation burst occurs, there is not much iron and sulfur left in the local solution since the solution has to be extremely supersaturated or enriched for pyrite to nucleate in the first place. Thus most of the iron and sulfur is removed and there is not enough left in solution for the crystals to grow substantially. This is another key point: the supply of new dissolved iron and sulfur to the growing pyrite crystals is then less than the rate of growth of the crystals. In fact, in most cases no further iron and sulfur reaches the framboid. This situation is achieved in environments such as fine-grained sediments, like muds, where flow of solutions is limited. This limited flow also allows the high levels of dissolved iron and sulfur to build up to unstable levels in the first place. Combining these hydrodynamic characteristics of the environment with the chemical requirements has enabled us to synthesize pyrite framboids in the laboratory, although we have not yet gained the strict control to produce organized framboids to order.

Today, some 10^{12} (a thousand billion) pyrite framboids are being formed every second.[18] As you read this sentence, there have been a thousand billion nucleation starbursts in sediments and another thousand billion tiny pyrite raspberries have been formed. One thing we know about Mother Nature: she has an unusual predilection for pyrite framboids.

Designer Pyrite

The idea that the form of pyrite crystals is determined by the environmental conditions in which they grow explains the infinite variety of shapes shown by pyrite in nature. This has led my research group to the concept of designing pyrite crystals with various shapes. This is not an idle pastime but has real applications in the Earth and environmental sciences and in materials science. For example, if we understood what controlled the shape of a natural pyrite crystal, we would know what the environment was like when the crystal was formed millions of years ago.

The key parameters determining the shape of a pyrite crystal include the degree of supersaturation, which in turn is linked to temperature and pH, and the hydrodynamics of the crystal growth environment—that is, how fast new iron and sulfur is supplied in the solutions. This latter has been largely overlooked in the literature but is key to determining and interpreting pyrite crystal form. In a perfect system where the concentrations of dissolved iron and sulfur are constant, the crystal form will be maintained throughout the growth process. But in most natural systems, these concentrations are limited because pyrite crystal growth is faster than the supply of new nutrients. Thus, as we have seen, in the simplest case octahedra grow until the supply of dissolved Fe and S is used up and then cubic faces develop.

Figure 4.17 shows a selection of the weird and wonderful pyrite forms we have made in our laboratory in Cardiff by varying the degree of supersaturation and the hydrodynamics of the system, such as stirring rate, which determines how fast new iron and sulfur can be supplied to the crystal. The syntheses were carried out at room temperatures in aqueous solutions. We were particularly proud of the donuts, although we do not

FIGURE 4.17. Designer pyrite: a selection of the forms of pyrite synthesized in the Cardiff laboratory: (a) simple cubes; (b) simple octahedra; (c) octahedron modified by cube; (d) rockets: prismatic cubes topped by octahedra; (e) donuts; (f) skyscrapers. The scales are in micrometers (μm).

know of a natural pyrite crystalline equivalent.[19] All the others have natural equivalents—usually rather larger[20]—but the same principles apply.

The synthetic pyrite forms explain many of the myriad of forms of pyrite observed in nature. As noted earlier, Steno's Law states that the angles between crystal faces are constant. Thus a pyrite crystal consisting solely of cubic faces does not necessarily produce a cube of pyrite: it could produce a square prism with the sides much longer than the top and bottom. In fact, because of the low symmetry of the arrangements of atoms within pyrite, different cubic faces have different surface energies. The surface energy is related to the ease with which more pyrite can nucleate on the pyrite surface from the iron and sulfur in solution. Then the rate of growth of the crystal faces is different on each face and square prisms are more probable results than perfect cubes. As previously mentioned, in the extreme case this feature may result in the formation of pyrite wires, where the side faces of the cubes are many times longer than the top and bottom faces.

This explains the formation of balls of radiating pyrite crystals. Balls of pyrite are commonly found in limestone and chalk where they are produced from the sulfur and iron in groundwater. Some of these spherical nodules consist of radiating, needle-like crystals and are commonly mistakenly identified as marcasite, the dimorph of pyrite. Marcasite crystallizes with orthorhombic symmetry, and thus it is supposed that long crystals of iron sulfide represent marcasite rather than pyrite, as discussed with respect to strike-a-lights in Chapter 2 and illustrated in Figure 3.3 in Chapter 3. In fact most of these nodules are pyrite rather than marcasite.[21] The individual pyrite crystals have simply grown into elongated forms that radiate from a center.

The overall process is illustrated in the synthetic forms in Figure 4.18. The first crystals form a cruciform-like habit as the top faces grow far more rapidly than the faces at the side of the cube. This can be enhanced in less agitated solutions where the only way that new nutrients can reach the sides of the crystal is by diffusion, whereas the tips extend into more agitated regions of the solution where new nutrients are continually being supplied. Ultimately a spherical hedgehog of radiating pyrite crystals is produced. The pyrite nodule shown in Figure 3.3 in Chapter 3 was formed in a similar manner.

We can see that by varying the concentrations of dissolved iron and sulfur and the hydrodynamics of the solution a vast array of forms of pyrite crystals can be produced. This explains extraordinary plethora of pyrite

FIGURE 4.18. The formation of radiating balls of pyrite: (a) skeletal pyrite prisms grow outward from a central cube; (b) prismatic pyrite crystals grow outward in intermediate space; (c) the process continues until the whole spherical pyrite crystal growth region is filled with radiating pyrite crystals. The scales are in micrometers (μm).

forms illustrated by Henckel in 1725 (Figure 4.2) and how it comes about that, as mentioned in the introduction to this chapter, a chemically simple mineral such as pyrite may exhibit the greatest variation in natural crystal forms in the mineral kingdom.

Notes

1. See L. Thorndike. 1923. *A History of Magic and Experimental Science: Vol. 12. The Seventeenth Century* (Whitefish, MT: Kessinger), 708pp.
2. L. Thorndike. 1923. *A History of Magic and Experimental Science.*
3. R.J. Haüy. 1801. *Traite de Mineralogie* (Paris: Louis).
4. J.V. Baez. 2009. *Who Discovered the Icosahedron?* Paper presented at the American Mathematical Society Fall Western Sectional Meeting. Riverside, CA. http://math.ucr.edu/home/baez/icosahedron.
5. E.S. Fedorov. 1891. *"Симмтрія правильныхъ системъ фигуръ"* ("Simmetriia pravil'nykh sistem figure") [The symmetry of regular systems of figures], *Zapiski Imperatorskogo S. Petersburgskogo Mineralogichesgo Obshchestva* [Proceedings of the Imperial St. Petersburg Mineralogical Society], series 2, vol. 28, pp. 1–146. English translation: David and Katherine Harker (trans.), *Symmetry of Crystals, American Crystallographic Association Monograph No. 7* (Buffalo, N.Y.: American Crystallographic Association, 1971), pp. 50–131. This was independently confirmed by the German mathematician, Arthur Schoenflies. 1891. *Kristallsysteme und Kristallstruktur* (Leipzig: Druck und Verlag von B.G. Teubner), 622pp.
6. This is a classical explanation of diffraction. In the postclassical, quantum world, a number of other explanations have been proposed. These involve the idea that the photon randomly chooses a path with a probability determined by Schrödinger's wave function. The probability that the photon will be scattered

in any particular direction is given by the square of the amplitude of the sum of the scattered waves. Then the intensity of a diffraction line or spot, which is a measure of the number of photons at that spot, is the square root of the intensity, and this can be used to compute the electron density.

7. In the race to publish his discovery, Laue published two short notices in a local journal, *Sitzungberichte de Königlicht Bayerischen Akademie der Wissenschaften Mathematisch-physikalische Klasse*. The first was in June 1912 with his collaborators W. Friedrich and P. Knipping (Interferenzerscheinungen bei Röntgenstrahlen), pp. 303–304, and the second was a solo effort in July 1912 (Eine quantitative Prüfung der Theorie für die Interferenz-Erscheinungen bei Röntgenstrahlen), pp. 368–369.

8. W.H. Bragg and W.L. Bragg. 1913. The reflection of X-rays by crystals. *Proceedings of the Royal Society of London, Series A*, 88:428–438. W.L. Bragg. 1913. The structure of some crystals as indicated by their diffraction of X-rays. *Proceedings of the Royal Society of London, Series A*, 89:249–277. W.L. Bragg. 1914. The analysis of crystals by the X-ray spectrometer. *Proceedings of the Royal Society of London, Series A*, 89:468–489.

9. Although it looks a lot more than four on Figure 4.7, it can be explained by thinking about the Fe atoms. There are eight Fe atoms at the corner of the cells, but since the crystal is a continuous lattice of these unit cells, each Fe atom is shared by eight adjacent unit cells. So the number of corner Fe atoms assigned to any individual unit cell is $8/8 = 1$ atom. Each of the six faces of the pyrite unit cell has a Fe atom in the center. These atoms are shared between two adjacent cells, so the total number of Fe atoms assigned to any individual unit cell is $6/2 = 3$ atoms. This gives the total number of Fe atoms in the pyrite unit cell as four, and since each is balanced by two S atoms in the pyrite formula, the number of FeS_2 molecules in a pyrite unit cell is four.

10. J. Alonso Azcárate, A.J. Boyce, S.H. Bottrell, C.I. Macaulay, M. Rodas, A.E. Fallick, and J.R. Mas. 1999. Development and use of in situ laser sulfur isotope analyses for pyrite-anhydrite geothermometry: An example from pyrite deposits of the Cameros Basin, NE Spain. *Geochimica et Cosmochimica Acta*, 63:509–513. These large pyrite crystals from the Rioja district were used to demonstrate some of the early applications of a new system for analyzing microscopic variations of sulfur isotopes in minerals using a laser ablation technique.

11. For example, I.K. Bonev, M. Reiche, and M. Marinov. 1985. Morphology, perfection and growth of natural pyrite whiskers and thin platelets. *Physics and Chemistry of Minerals*, 12:223–232.

12. The key idea here is that the pairs of sulfur atoms need to be in a pyritohedral arrangement in order to provide the lower symmetry of the pyrite cubes, since the pyritohedron has only twofold symmetry whereas all the rest of the cubic forms have fourfold symmetry.

13. D.N. Marshall. 1977. Carved stone balls. *Proceedings of the Society of Antiquaries of Scotland*, 108:40–72. See also L. Le Bruyn. 2012. Scottish solids, final(?) comments. www.neverendbooks.org. Le Bruyn delicately phrases his self-criticism of his earlier work in which he claimed that these ancient Scots had discovered the Platonic solids, as due to poor scholarship.

14. Pentagons have fivefold symmetry, which means that any interior angle is 360/5°.The problem with them is that 360/5 is 108 and three pentagons would produce a maximum angle of 324° (leaving a 36° gap) and four pentagons overlap (4 × 108 = 432°).

15. I. Sunagawa. 1957. Variation in the crystal habit of pyrite. *Geological Survey of Japan Report*, 175:1–47. Sunagawa's work was qualitatively confirmed by the experiments of J.B. Murowchick and H.L. Barnes. 1987. Effects of temperature and degree of saturation on pyrite morphology. *American Mineralogist*, 72:1241–1350; and further developed in D. Rickard. 2012. *Sulfidic Sediments and Sedimentary Rocks* (Amsterdam: Elsevier, 801pp.), where the importance of hydrodynamics in the development of pyrite habits was first demonstrated.

16. They were named by G.W. Rust. 1931. Colloidal primary copper ores at Cornwall Mines, Southwestern Missouri. *The Journal of Geology*, 43:398–426.

17. Counts have shown 100,000 pyrite framboids per gram of dry sediment, which amounts to just 0.2% pyrite by weight. There are about 6×10^{21} g of pyrite buried in sedimentary rocks. If we assume that only 10% of this is framboidal, this would mean between 10^{29} and 10^{30} framboids with average diameters between 0.01 and 0.02 mm.

18. The current best estimate for the net rate of pyrite formation in sediments is 10^{14} g pyrite per year. If 10% of this is framboids, then this gives around 10^{12} framboids per second. This is a minimum figure, since most of the pyrite formed in sediments is reoxidized immediately and pyrite framboids are also being formed in volcanic-related processes.

19. There is a natural, microscopic texture in some pyrite ore deposits called *atoll texture* that is about the right size and may sometimes represent random sections through pyrite donuts.

20. The size of the crystals is basically a matter of time. Mother Nature has far more time in her natural laboratory. For example, a rate of 1 mm of new sediment per year is a fairly rapid deposition rate for the muds where pyrite formation occurs most readily. But even a year is a long time for a laboratory experiment. So most of our synthetic pyrite crystals are small. Larger crystals can be made at high temperatures using technologies not found in nature.

21. Maurice Huggins determined the marcasite structure in 1922: M.L. Huggins. 1922. The crustal structures of marcasite (FeS_2), arsenopyrite ($FeAs_S$) and loellingite ($FeAs_2$). *Physical Reviews*, 19:369–373. Buerger proved the structure in 1931 and noted with some typical Buergeresque asperity that Huggins' structure was correct even though his crystallography was entirely wrong: M.J. Buerger.

1931. The crystal structure of marcasite. *American Mineralogist*, 16:361–395. Thus radiating pyrite nodules are often called *marcasite* because of the crystal habit. More often than not they turn out to be pyrite, as shown by F.A. Banister. 1932. The distinction of pyrite from marcasite in nodular growths. *Mineralogical Magazine*, 23:179–187, who applied the X-ray diffraction technique to distinguish between pyrite and marcasite in pyrite nodules of the type used in the ancient strike-a-lights. He found that all those he analyzed from the chalk deposits of southern England were pyrite.

5

Hell and Black Smokers

MOST OF THE important metal ores in medieval and ancient times were pyrite-rich sulfides. These pyrite-rich ores were a major source of a suite of valuable commodities such as sulfur, arsenic, copper, lead, zinc, and nickel, as well as some gold and silver. This is why in 1725 Henckel could devote a 1,000-page volume to pyrites, *sensu lato*. Because of its relative abundance, its potential economic importance, and its exotic composition compared with the rock-forming minerals, pyrite has played a key role through the ages in developing ideas of how minerals and ore deposits form. During the last century, pyrite became an even more important mineral in discussions of ore genesis because it is also a key component of sediments. This led to conflicting theories of ore genesis, in which the ore minerals were formed in the sediments or introduced later, often by processes related to volcanism. The conflict between adherents of these theories continues to this day.

Ancient Ideas of Mineral Formation

Pyrite constituted a key, but sometimes uncomfortable, mineral in ancient theories of mineral formation. It was relatively common and often economically important. However, it contained sulfur as a key constituent and this contrasted it to many other common minerals and rocks in that this meant that pyrite could be changed by heating. Heating released sulfur from pyrite, leaving a residue of stony slag. The ancients also recognized sulfur as a special material since it occurred in solid, liquid, and gaseous form, rather like water. Any theory of mineral formation needed to explain how this protean element got into pyrite. This problem was compounded by the fact, discussed in Chapter 3, that for some unknown reason the ancients did not know that pyrite contained iron.

Ancient theories of mineral formation divide into three categories: (a) the Genesis theory: that all minerals were formed by God during the creation of the Earth; (b) the Aristotelian theory: that all minerals were formed at depth in the Earth through the interactions of the four basic elements; and (c) the Alchemical theory: that minerals were formed from combinations of mercury and sulfur.

The Genesis theory requires little comment except to note that people who proposed alternative explanations during the late Middle Ages in Europe needed to be strong characters or clever theologians. Pyrite presented a particular problem to the practicing mine geologist since it was observed forming or was proven to have formed very recently. In 1556 Agricola dismissed the Genesis theory as the "opinion of the vulgar" (i.e., ordinary, uneducated people).[1]

The Aristotelian theory was based on the four elements: earth, water, air, and fire. These were never found pure, and they were endowed with certain fundamental properties: dryness and dampness and heat and cold.[2] The elements were subject to transmutations by the action of the properties. For instance, heating removed the cold from water and produced air (i.e., steam). From the point of view of the origins of minerals such as pyrite, precipitation occurred when dampness—that is, water—was removed. This reflected the commonly observed phenomena of evaporation and distillation, which produced other minerals. However, the idea that pyrite could be formed by precipitation is interesting since no experiments were done to show this nor were there any descriptions of pyrite being formed in an aqueous environment. It can only have been a hypothesis derived from the application of Aristotelian principles.

In the Aristotelian theory, the transmutation of the elements within the Earth produced two types of exhalation: a gaseous emanation associated with fire produced stones and steam associated with water produced metals. Aristotle's pupil and successor Theophrastus[3] elaborated and clarified these ideas. The metals melted when heated and therefore must be, in large part, water and, like water, they solidified with cold. Therefore, metals were cold and damp. Stones, by contrast, solidified with heat and did not appear to melt; therefore they were dry and hot and must be, in large part, earth. So ore deposits were formed within the Earth by the transmutation of the four elements on mixing, including mixing with water.

Avicenna progressed these Aristotelian ideas into a relatively straightforward concept of mineral and rock formation: minerals were formed either by solidification from a liquid or by the aggregation of solids. Solidification from a liquid involved various processes that can be observed experimentally, such as crystallization and distillation. The process is equivalent to the Aristotelian removal of water. Aggregation of solids is the equivalent of lithification, the process whereby a loose aggregate of sand or clay gains become indurated to form a rock.

Although the language is widely divorced from modern concepts, we can see how the ideas of a practical mining geologist like Agricola developed from Aristotelian ideas. However, the pyrites constituted a problem in applying the Aristotelian classification. Agricola knew that heating pyrites produced sulfur but that metals, such as gold and silver, copper, and zinc, could be extracted from it, leaving behind a slag, which was stone-like. Therefore, Agricola classed the pyrites as mixed materials, being combinations of stone, sulfur, and metal.

The third strand of medieval thinking about ore deposits was provided by a group we loosely refer to as alchemists. These included people with ideas closely related to astrology. Albertus Magnus, whom we met in Chapter 3, was particularly concerned with the astrologic aspects of alchemy. Each metal was related directly to a planet or, as with gold, directly to the Sun. Since there were five planets other than Earth, plus the Sun and Moon, these correlated with the seven classic metals: gold, silver, copper, iron, mercury, tin, and lead. The astrologic idea was then that the occurrence of each metal ore was somehow due to the action of the relevant planets in favorable conjunctions, by drawing up vapors from the depths of the Earth for example. The discovery of new metals—such as bismuth originally recorded in the *Nützlich Bergbüchlin* of 1520[4]—cast a pall over these astrologic ideas.

However, alchemy continued to thrive into the 17th century and alchemical ideas influenced thinking about how pyritiferous mineral deposits formed. The Pseudo-Geber idea that all metals were mixtures of sulfur and mercury encouraged the thesis that these aspects of Aristotle's *prima materia* within the Earth intermixed in various proportions to produce the ore minerals like pyrite. Albertus Magnus promoted these ideas and added salt to the mix as another aspect of the *prima materia*. His reasoning was that evaporating saltwater produced salt and that this aspect of transmutation needed to be included in the ore-forming mix.

The basic alchemical thesis developed from the admixture of Eastern alchemical theories with Greek philosophy in the 1st century CE, and it was especially centered in Alexandria in Egypt. Aspects of Aristotle's four elements were admixed with Chinese ideas of mercury being a key secret component of metals. The general alchemical thesis for the origin of metallic ores was that less noble metals grew deep in the Earth and they changed over time to more noble metals. So pyrite was on its way to gold but had not yet reached this degree of purity. Indeed, if a vein ran out or diminished, the miners sometimes sealed it off and allowed it to recover so that more metals grew. This idea is diametrically opposite to the Genesis theory, which stated that all metals were formed at one time. This Egyptian-based Greek alchemy was preserved and developed by medieval Arab writers such as Avicenna. It reached western Europe in the 11th century, where it was recycled into Latin by Albertus Magnus.

Pyrite and Volcanoes

The frontispiece of several medieval volumes on mining and economic minerals features an active volcano. Figure 5.1 shows the frontispiece of Henckel's massive work on pyrites. A shaft, complete with winding gear, is situated high up on the flanks of an active volcano. A miner is wheeling the extracted ore down the path in a wooden wheelbarrow. The ore is taken to three huts where sulfur, arsenic, and vitriol (iron sulfate in sulfuric acid) are variously extracted according to the astrologic symbol for the substance that adorns the hut. These products are packed up and taken on horse-drawn sledges to the port, where they are loaded into sailing ships. In the foreground the owner discusses some matter with the chief mining engineer.

The volcano is presented as the source of the pyritic ores the miners are exploiting. The curious thing here is that no active volcanoes occurred in Germany, the Czech Republic, or Poland at the time of Henckel's work nor had there been any for millions of years. The artist could not have actually visited a mine because, although they were often situated on mountainsides, they were rarely near a port and never in an active volcano. Interestingly, this relationship between ore minerals and volcanoes is featured in the medieval literature of countries where active volcanoes are unknown in historical times. The ore–volcano relationship appears to be culture-independent: it is figured in the Chinese and Indian as well as European literature.

FIGURE 5.1. Frontispiece of Henckel's (1725) volume, *Pyritologia*. The illustration depicts the various activities associated with contemporary mineral exploitation and includes an active volcano in the background.

The idea that minerals such as pyrite were originally formed in volcanoes was perhaps most strongly expressed by Benoit de Maillet in 1721:[5]

> It is to these volcanoes, whether active or not, that we owe all minerals and metals, gold, silver, copper, lead, tin, sulfur, alum, vitriol and quicksilver, which their fire has at first deposited on the sides of the vents opened up by their flames, like the soot of the wood and coal that we burn is deposited on ours.

Again this statement flies in the face of the actual mining experience of de Maillet, even though he was a much-traveled French diplomat who had served in Cairo, Livorno, the Levant, and the Barbary Coast. His own experience of mines to the east of his birthplace in Lorraine must have shown him that all the metals he listed were commonly found in rocks that were not formed in volcanoes whether active or ancient. It seems as though the great natural drama that is represented by an active volcano, and which travelers such as de Maillet certainly observed, seared themselves into the consciousness to such a degree that highly rational people could reach the same conclusion as de Maillet in the face of overwhelming, everyday evidence to the contrary.

The cultural association of sulfur with volcanoes is long lived and gave rise to the medieval association of sulfur with Hell. Various volcanoes have been identified as the mouth of Hell. One of the most famous is Hekla in Iceland, which has been among the most active volcanoes in the world during the last millennium with eruptions lasting as long as six years. Hekla is infamous for the noises heard during eruptions, which sound like the groans and wailings of the damned (Figure 5.2). These stories spread throughout Europe particularly after the eruption of 1104 CE, and the belief that Hekla was a gateway to Hell lasted well into the 1800s.

I visited White Island, a marine volcano situated some 40 km off the north coast of New Zealand, during its active period in 1992. You may ask, how do you visit an active volcano? The answer is: quickly. You literally run around it, keeping to the narrow paths that indicate where people had been before. Straying from the paths is dangerous. A Finnish colleague lost a foot in the molten lava just off a path when we were visiting a volcano in northeast Iceland. The black lava had cooled enough to form a crust but not enough to support his weight. White Island is interesting since it is sulfur-rich. Indeed, various attempts had been made to mine sulfur until an eruption killed all the workers in 1914. Figure 5.3 shows the White Island volcanic crater from the inside. You can see the yellow sulfur deposits in the crater wall. The white vapor is a sulfurous steam and is quite poisonous. So you have to be careful in these environments to make sure the wind keeps the fumes away. Mind you, if the wind changes suddenly...

Although sulfur and pyrite are associated with volcanic activity, neither actually forms commonly in the volcanic rock itself. The iron sulfide in solidified lava is usually pyrrhotite, an iron monosulfide, rather than pyrite. The pyrite and sulfur are mostly found in the *fumaroles* in and around the volcano.

FIGURE 5.2. Looking into the mouth of Hell. The Hekla eruption of 1980. The lava is at a temperature of around 1100°C, which compares with a hot oven at 200°C and molten lead at 327°C (see color plate).

Fumaroles are gaseous emanations from volcanoes. They might be described in Aristotelian terms as geothermal springs where the water has been boiled off. The ancients in the Middle East and around the Mediterranean were familiar with active volcanoes and their products. In preclassical times, the nearest active volcano to Mesopotamia was Nemrut in Anatolia, and the eruption of the Anatolian volcano, Hasan Dag, in 6200 BCE is the oldest recorded volcanic eruption. The nearest volcanoes to Egypt were in northwestern Saudi Arabia, the land of the Midianites, where Moses spent 40 years. Sulfurous fumaroles occur on the upper slopes of Mount Damavant, a mostly dormant volcano that features strongly in ancient Persian folklore. In Greece, the Methana, Milos, Nisyros, and Santorini volcanoes have been active in historical times. In Italy, Stromboli, Etna, and Vesuvius have erupted in the past 100 years but Pantelleria, Vulcano, Campi Flegrei, Ischia, Larderello, Lipari, and Vulsini all erupted in historical times.

FIGURE 5.3. Inside the active volcanic crater at White Island, showing the yellowish sulfur deposits in the crater wall and the sulfurous steam.

The observations of sulfur and pyrites being formed in volcanic fumaroles led to the generally accepted medieval and early Renaissance idea that pyrite ores were formed in volcanoes. As we have seen, this was accepted in mining areas where volcanoes were unknown. Indeed, even the rocks in which the pyrite was mined were often not volcanic but lithified sediments. It seems that the widespread belief that pyrite and other sulfide minerals were formed by volcanic activity was at best a case of observations being manipulated to fit a hypothesis. At worse it was a peculiarity of the human consciousness that a conclusion could be reached about mineral formation that entirely ignored observations of the natural world.

Sulfur and pyrite are produced around the volcanoes. What is less obvious is the association of valuable metals with volcanoes as indicated in the Henckel and Becher drawings. It may well be that this is pure Aristotle: metals grew deep in the Earth and were brought to the surface in various states of maturity by volcanoes. In other words, the hypothesis of how metals formed determined the thinking about their formation rather than the actual observations about their occurrence. The observations of pyrite formation around volcanoes would have contributed to this idea since, even though the pyrite did not necessarily contain valuable metals, it proved that volcanoes could produce sulfide minerals. However, as discussed earlier, the idea of an association between metals and volcanoes appears to be culture independent and therefore Aristotle's ideas cannot be the ultimate source. It possibly suggests even more ancient roots before the development of distinct regional cultures.

Pyrite and the Problem of Geologic Time

Pyrite played an important role in thinking about the age of the Earth because it was economically important and stood out as golden crystals in the duller rocks. It often appeared to have been formed after the rocks that contained it. Ideas of how and when pyrite formed were therefore inevitable consequences of any theory of the formation of the Earth.

It was not until the middle of 20th century that the age of the Earth was finally fixed at 4,570 million years by the analysis of radiogenic isotopes. The concept of deep time, however, had been one of the main revolutionary ideas of human culture in the early years of the 19th century. It had not only provided the basis for the Earth and planetary sciences but had also provided the framework in which Darwin could develop his theory of evolution. The evidence that the Earth was very old was originally found by studies of the relationships between rock strata. Even so, the debate about the age of the Earth was a relatively recent one in historical terms. It developed after the Protestant revolution as a consequence of the literal reading of the Bible suggesting that the Earth was only 6,000 years old. Prior to that time, the ancients had no problem with the Earth being immensely old, and Aristotle wrote that he could see no beginning or end, a sentiment later echoed by the Father of Geology, James Hutton, in 1799.

It was originally assumed that pyrite and its associated minerals were forged in the great heat at depth in the Earth and then brought to the surface by volcanoes. It is then interesting to ask when this occurred, since volcanoes are not situated in many of the areas where the minerals were actually mined. In the post-Protestant, Christian hegemony this was often assumed to have been in a period before the Great Flood, or during the Great Flood, or during the actual creation of the Earth. Henckel was quite wary about this in 1725. He stated that "it is probable that the pyrites did not all originally exist from the creation, but that most of them have been generated at different and successive periods; which generation will, in all likelihood, continue to the end of all things."

However, he also concluded that much of the pyrites found its present location and was formed as a result of a cataclysmic, water-dominated event, which he interpreted as the Deluge. In a prescient paragraph he also noted pyrite forming today on wood as well as in "periwinkles, muscles and the like shells."[6]

Of course Aristotle and his colleagues from the classical period did not have the proscriptions about the formation of the Earth that constrained the philosopher-scientists of the Common Era, particularly during the period following the Protestant revolution. They did not need to believe that the Earth was created in seven days and that it was just a few thousand years old. Herodotus (484–425 BCE), for example, estimated the time taken for silting up the Nile delta and thought that this would have taken at least 10,000 years. So we assume that Herodotus must have accepted that the Earth was far older than this.[7] Neither were the ancients necessarily constrained by the Great Flood and the need to interpret geologic and mineralogic observations as evidence for this event. Aristotle followed Xenophanes (570–480 BCE) in identifying fossils as petrified animals and accepted that the positions of the land and sea had changed with time. He wrote in *Meteorologica*[8], his geologic treatise: "But the whole vital process of the Earth takes place so gradually and in periods of time which are so immense compared with the length of our life, that these changes are not observed." Aristotle's view on the rate of Earth processes should be considered in the context of his idea that the Earth is eternal and has no beginning or end.

Abu Rayhan al-Biruni (973–1048 CE), whose mineral list we mentioned in Chapter 3, proposed that the ocean once covered India. Avicenna may have first proposed the law of superposition of strata, which states the younger strata occur above older ones, some 500 years before Steno.[9] The importance of Avicenna's insight is that as you look at a pile of rocks, you are looking back in time—and there was enough time for mountains to be formed and eroded and the sediments piled up. Avicenna, as a devoted follower, concurred with Aristotle that the Earth had no beginning or end. The Chinese naturalist Shen Kuo also recognized that the Earth was very old. In his *Dream Pool Essays* (夢溪筆談, Mengxi Bitan) of 1088 CE he described marine fossils found inland and proposed a theory of the formation of land based on the erosion of mountains and silt deposition over extremely long time periods.

During and after the Protestant revolution European scientists struggled with the consanguinity of their observations with contemporary religious dogma. Pyrite played an interesting role in this debate. It was observed forming in organic matter, hot springs, and volcanoes. It was also observed in fossils like shells and plants. In the ultimate hokum of the Christian hegemony, pyritized ammonites were proposed to have been placed in rocks by Satan in order to mislead the godless. The formation of veins of pyrite were all related to the waters of the Deluge circulating in fractures or erosion channels in the rocks. Otherwise, all the pyrite was formed when the Earth was created and had not changed since. This became untenable with the obvious observable oxidation and destruction of pyrite, the ongoing formation of pyrite within thermal springs, and the formation of pyrite in contemporary organic matter. It is no wonder that the medieval Schoolmen thrust their heads into the sand and thought that actual observations of nature were confusing.

The problem of when a mineral like pyrite was formed was not one that worried these ancient writers. Pyrite was formed in the past and was being formed today. The amount of time available for pyrite formation was immense. In this sense, the ideas of the ancients about time were similar to recent ideas about deep time: the Earth is 4,570 million years old and therefore there is an immense amount of time available for pyrite formation.

Pyrite and the Origin of the Earth's Interior Heat

The idea that the interior of the Earth is hot has been around since ancient times. It is easy to see why. Volcanoes appear like holes in the Earth's surface, providing a view of the interior. And this is so hot that rock melts. More subtly, although ancient miners would have known that as they went further down into the Earth it became warmer, it appears not to have been documented until 1664 by the German Jesuit priest Athanasius Kircher.[10] The problem for the philosophers was: Where does this interior heat of the Earth come from? The ancients could not melt rock themselves, and thus the heat produced in the Earth was greater than any human-made fires. Pyrite was thought to have played a central role in generating the heat of the Earth for several hundred years. Indeed, the pyrite theory was one of the first scientific theories about the origin of Earth heat.

Pythagoras of Samos (540–510 BCE) was one of the earliest writers to propose that there was a central fire within the Earth, although he may have inherited these ideas from Thales of Miletus (624–546 BCE).[11] Anaxagoras of Clazomenae (500–428 BCE) thought that the *aether*, Aristotle's fifth, quint-essential element that filled interplanetary and interstellar space, sank into the hollow interior of the Earth where it mixed with vapors, causing fire. The hot air rose toward the surface and, where its passage was blocked, was forcefully expelled, causing earthquakes and volcanoes. The idea that volca-nic eruptions were caused by air trapped under great pressure within the Earth became a basic cornerstone of scientific thinking through a millen-nium. Aristotle in particular developed the theory of winds within the Earth that produced volcanic eruptions. The elements of air and water trapped within the Earth were heated and expanded upward to cause hot, explosive volcanism.

Paracelsus (1493–1541) was the great alchemist of the late Middle Ages in Europe. He was actually an Austrian physician born Philippus Aureolus Theophrastus Bombast von Hohenheim and had some knowledge of mining and metallurgy in the Tyrol. Paracelsus believed that there was a great fire in the center of the Earth that produced dense clouds of vapors. These vapors condensed to form ore deposits. The evidence appeared to be active volcanoes and hot springs where clouds of steam and ashes were ejected during eruptions and the surrounding ground was hot, simmer-ing and fuming with molten sulfur even between eruptions. Paracelsus was a direct inheritor of the classic Greek thinkers on the nature of the Earth and the origin of volcanoes.

Various authors in the Middle Ages adopted combinations of these ideas to suggest theories of ore formation (e.g., Figure 5.4). For example, the only work on mining geology to precede Agricola in the medieval period was *Ein Nützliche Bergbüchlin* of 1520 mentioned earlier. This presented a confused account of the origin of ore deposits that combined both astro-logic and alchemical ideas basically following Albertus Magnus. Ristoro d'Arezzo concluded in 1292 CE that the interior of the Earth was very hot and followed the minority view of the Greek philosopher Empedocles of Agrigentum (495–435 BCE). Empedocles, who, as we discussed in Chapter 3, insisted that there were just four elements, thought the Earth had a molten center and that volcanoes erupted through the rise of molten rock to the surface.

In the 16th and 17th centuries the heritage of the ancient alchemists led many to believe that exothermic or heat-producing chemical reactions between sulfur and iron or between iron and pyrite were responsible for providing the heat within the Earth. The idea was based on the observations of the ancients that sulfur occurred abundantly in and around volcanic craters and burned with a sputtering blue flame, producing strong-smelling, sulfurous gases. So the reason for the idea that volcanoes were formed by the combustion of sulfur buried in the Earth is understandable. Similarly, hot springs are often sulfurous in nature. These sulfur-bearing waters have long been supposed to be health-giving if not a cure to a long list of ills. My own experience of these sulfurous waters is

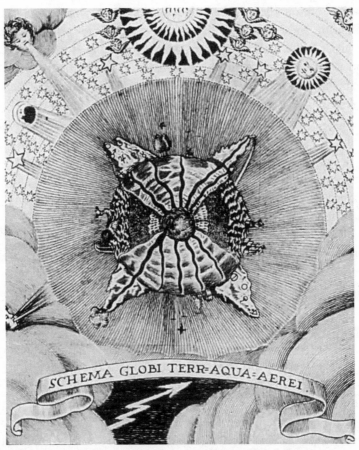

FIGURE 5.4. A cross section of the Earth according to Johann Joachim Becher in 1719 showing the molten center feeding volcanoes that contain metals and sulfur, represented as astrologic symbols. From J.J. Becher. 1719. *Opuscula Chymica rariora,* edited by F. Roth-Scholtzio (Norimbergae: Apud haeredes Joh. Dan. Tauberi), p. 74.

that they taste even worse than they smell, which may be why they were thought to be good for you. However, they provided evidence to support the idea that sulfur combustion heated the groundwaters and generated hot springs.

This was not a trivial idea: authorities such as Isaac Newton also subscribed to the idea of a chemical origin of the Earth's heat. This derived primarily from his secret alchemical experiments, which he did in his shed in the garden behind his rooms in Trinity College, Cambridge. He wrote in his *Opticks* of 1704:[12]

> And even the gross body of sulphur powdered, and with an equal weight of iron filings and a little water made into a paste, acts upon the iron, and in five or six hours grows too hot to be touched and emits a flame. And by these experiments compared with the great quantity of sulphur with which the earth abounds, and the warmth of the interior parts if the earth and hot springs and burning mountains, and with dumps, mineral coruscations, earthquakes, hot suffocating exhalations, hurricanes, and spouts, we may learn that sulphureous steams abound on the bowels of the earth and ferment with minerals, and sometimes take fire with sudden coruscation and explosion, and if pent up in subterraneous caverns burst the caverns and with a great shaking of the earth as in springing of a mine.

Newton appears to have been looking at what we now know is the oxidative dissolution of pyrite or, at least, an iron sulfide. This reaction is highly exothermic: it generates about 10,000 J of chemical energy for every gram of pyrite oxidized. This is theoretically enough to keep a 13-watt energy-saving light bulb going for more than eight days. The amount of heat actually produced depends on factors such as the rate of reaction and the conditions pervading the system at the time. However, the heat produced is often sufficient to cause spontaneous combustion.

The idea of reactions involving pyrite being responsible for the Earth's interior heat was widespread in pre- and early Renaissance science. Martin Lister, for example, was an advocate. In 1682 he wrote that volcanic eruptions were caused by the heat generated by pyrite oxidation. In 1700 Nicolas Lémery buried wet sulfur and iron in the ground and found that within a few hours they had reacted and a flame erupted from the surface together with some explosions.[13] This appears to be one of the earliest expressly geologic experiments.

These oxidative dissolution processes are so intense in some pyrite ore deposits that large quantities of heat are produced and the pyrite appears to burn—although the actual combustible material is usually associated sulfur, organic matter, or bitumen. One of the most famous cases is the Stora Kopperberg (Great Copper Mountain) mine at Falun in Sweden (Figure 5.5). This mine produced copper from copper pyrites from around 1100 CE to 1982. It was the world's largest copper mine in the 16th and 17th centuries and the source of most of Sweden's wealth. There was a major collapse of the ancient mine workings in 1687, which produced a pit 100 m deep and up to 400 m across. This allowed water and oxygen to penetrate a mass of broken pyrite. The result has been the production of an enormous mass of iron oxide, which is sold as Falu Red, the paint providing the distinctive red color of the characteristic red and white Swedish country houses. The iron ocher is also a wood preservative. The oxidation products include a variety of exotically colored sulfate minerals as shown in Figure 5.5. These sulfates are the vitriols and alums of the older

FIGURE 5.5. The effect of 300 years oxidation in a massive pyrite ore. A collapsed stope in the Falun mine, Sweden, showing iron, copper, and calcium, magnesium, zinc, and lead sulfate stalactites in a mass of iron oxide ocher. The temperature can reach 50°C here (see color plate). Photo Folke Schippel and Rolf Hallberg. From D. Rickard and R.O. Hallberg. 1973. De små gruvbetarna i Falun. *Svensk Naturvetenskap*, 102–107.

literature, which were key substances in ancient medicine and chemistry as discussed in Chapter 2. The exothermic oxidative dissolution reaction has been going on ever since, and Figure 5.5 shows about as far into the mine as we could penetrate: the temperature reaches 50°C even in this stope. The center of the old mine workings cannot be approached because of the heat. In 1973 my colleague, Rolf Hallberg, and I showed that the amount of heat being produced would be enough to light a small town like Falun.[14]

Other heat-producing chemical reactions were proposed, including the reaction between iron and sulfur to produce pyrite. This reaction was called upon to explain how the exothermic oxidation of pyrite could occur deep in the Earth where there was no air. The idea was that the reaction between sulfur and iron would first occur at depth and break up the surface layers so that the air could then penetrate to cause the more intense oxidative process.

In 1644 Descartes revolutionized thinking about the origin of the Earth's interior heat with his idea that the Earth was like a small star that was cooling from an original molten state.[15] The Earth's heat is then primordial and is what is left since the Earth was formed. However, the chemical theory of the origin of Earth's heat lasted for a considerable time, especially as an explanation for volcanoes. It was thought that the interior heat of the Earth was generally primordial and local chemical reactions produced excess heat that resulted in volcanic eruptions and hot springs. This idea was supported as late as the mid-20th century by that great but often wrong, English geophysicist Sir Harold Jeffreys. This was more than half a century after the discovery of radioactivity and the realization that most of the Earth's heat is generated by the breakdown of radioactive elements such as potassium, thorium, and uranium in the Earth. A small part, perhaps a fifth, is primordial.

It is interesting that pyrite had a leading role to play in the development of the ideas about the origin of the internal heat of the Earth and volcanoes. The chemical theory of Earth heat has long been discarded. However, in the absence of knowledge of radioactivity, pyrite provided a potential means to generate heat. The reactions of interest appear to have been twofold. First, the reaction of iron with sulfur is simply a recipe for synthesizing pyrite. This reaction also produces heat but not as spectacularly as the oxidation of pyrite. The second reaction involves the oxidation of pyrite, which can be so rapid as to result in spontaneous combustion. The mixtures of sulfur, iron, and water that the alchemists used to produce spectacular displays of chemical heat production essentially involved the formation of finely divided pyrite by the combination of iron and sulfur followed by its rapid oxidation Although these theories were to prove mistaken in the light of later discoveries, they provided a basis for understanding and predicting the role of pyrite oxidation in the environment today.

Pyrite in Geothermal Springs

One feature of Agricola's classic 1556 work on mining and metallurgy is that he described how minerals like pyrite occurred in nature. This contrasted with the philosophy of the medieval Schoolmen who thought that nature was so complex that actual observation would merely confuse matters. Agricola reported that pyrite occurred mostly not in volcanoes but in *veins* that crisscrossed the countryside (Figure 5.6). The word *vein* comes directly from the Latin *vena* and was perhaps first recorded by Pliny.[16] However, there is no explanation for why these mineral fissures were called veins.

FIGURE 5.6. One of Agricola's many diagrams of veins in *De re metallica*, 1556. This is taken from Book 3, which is mostly concerned with vein mineralization.

The veins were often in sedimentary rocks such as limestone and shales. These veins usually included the richest ores with the largest mineral crystals, and the metals were easier to extract with the restricted metallurgy available at that time. They were also easier to mine. Individual miners could follow veins underground and remove the ore with hand tools. In this type of mining, not much country rock had to be removed to get at the ore. Obviously, the more worthless country rock that needs to be removed per ton of ore decreases the effective value of the ore in dollars per ton.

Agricola described veins in some detail and concluded that these were basically fissures in the country rock that had been later filled with ore and gangue minerals. Gangue minerals are minerals in the vein that are not valuable, such as quartz and calcite. Agricola proposed that the minerals in the veins were precipitated from solutions, usually a mixture of heated groundwater and mineral-rich solutions emanating from depth in the Earth which he called fluids (*succus*). He described hot springs as the surface expression of veins.

The ideas of Agricola concerning veins were revolutionary at the time and not widely accepted until over 200 years later. They can be seen as significant extensions of contemporary ideas as expressed by Aristotle and, more practically, by the description by von Kalbe in his *Bergbüchlin*. The woodcuts of vein types presented by Agricola in *De re metallica* are eerily similar to those published earlier by von Kalbe, although more extensive. Agricola seems to have been more analytical than his predecessors, and his presentation of ore formation is relatively straightforward. They reflect our present understanding of vein formation, which is described as being a result of hydrothermal (i.e., hot-water) processes. The only negative aspect of Agricola's thesis was his insistence that the original fissures were caused by surface erosion, like caves. Although this sometimes happens in limestone terrains, most fissure veins are caused by tectonic processes that occur within the Earth. This would not be realized until the late 18th century.[17] Indeed, it was not until the late 20th century that the relationship between fluid pressure, fissuring, and vein formation was understood.[18]

Geothermal springs are the surface expressions of veins. They indicate where veins are forming in the Earth at present. The material that has not been precipitated in the veins debouches onto the surface in geothermal springs.

Geothermal springs occur in volcanic areas, such as the New Zealand springs shown in Figure 5.7. In these cases, the volcanic heat itself is

FIGURE 5.7. Sulfur-rich emanations, including pyrite as well as rarer minerals such as orpiment and gold, in the boiling geothermal springs of Champagne Pool, Waiotapu, New Zealand (see color plate).

heating the groundwaters. In Iceland one of my doctoral students, Peter Torssander, showed that the geothermal spring waters in the central areas of the island were mainly rainwater that had been reheated by the volcanic systems. By contrast, around the coast the main source of the spring water was seawater. The seawater, which is rich in sulfate, provided the sulfur in the geothermal waters.[19]

In 1846 Mount Hekla erupted in Iceland and the Danish government invited Robert Bunsen to investigate the eruption. Bunsen is more famous for the eponymous Bunsen burner, which is found in every chemical laboratory. However, he was also a keen field geologist. Bunsen studied the geothermal springs of Iceland and made several fundamental observations on pyrite formation in geothermal systems, including the fact that pyrite was formed through the reaction between polysulfides, chains of dissolved sulfur atoms, and iron.[20] In fact, these observations were virtually ignored for 150 years until I reported them in the international literature.[21]

However, many springs occur in areas where there is not and has never been any volcanic activity. The water in these springs is groundwater that has penetrated deep into the Earth's crust and been heated by the elevated temperatures that occur with depth.

For example, there are several hot springs near to where I am writing this, at Hotwells near Bristol, Taff's Well near Cardiff, and, most famously, at Bath. All these springs are sourced in Carboniferous limestone, a sedimentary rock deposited some 300 million years ago. The Bath springs are the most famous since the original Roman bath house is preserved. The Bath springs bubble out from three sites. The Sacred Spring produces over 1 million liters of water per day heated to 46°C and derives from groundwaters that penetrated at least 2.5 km into the Earth's crust. The water is ultimately sourced from rainfall on the nearby hills. The flow rate in Bath is relatively low compared with high-magnitude springs such as the Silver Springs in Florida, which produces almost 2 billion liters per day. The waters at Bath, as with many springs, are slightly sulfurous and have been and still are being used for medicinal purposes. The treatments include both bathing and internal digestion. The water at Bath has a disgusting taste so I suppose, as noted earlier, that it is very good for you.

Geothermal systems and hot springs explain many pyritic vein deposits. The waters are heated either by the geothermal gradient of the Earth or by volcanic sources. The sulfur is mainly leached from the rocks by the circulating hot waters, although there may be some components related to sulfurous gases evolved on molten rock cooling in volcanic areas. The iron and associated valuable metals are leached from the rocks by the hot waters and react with the sulfur and pyrite precipitates as the solutions cool. The hot waters are directed along fractures and the sulfides coat the fracture walls, forming veins. Mineral precipitation can block the fractures, leading to a build-up of pressure and successive cycles of fracturing, pressure release, temperature drop, mineral precipitation, and fracture blockage. The result is the classical banded veins of minerals as the cycles are repeated.

It seems obvious today that if geothermal springs occur on land then they are even more likely to be present in the oceans, which cover the greater part of the Earth's surface. The temperature and salinity anomalies caused by the warm geothermal brines of the Red Sea were reported first by the Russian ship *Vitaz* in the 1880s and then by the Swedish research vessel *Albatross* in 1948. The Swedish expedition did not discover the brines themselves because the chief geochemist and now doyen of the oceanographic research community, Gustaf Arrhenius, cut short his cruise to return home to his girlfriend. The temperature and salinity anomalies were thought to be caused by solar heating of these tropical waters. However, in 1964 the British research vessel *Discovery* reported extraordinary temperatures of 44°C, and the US research vessel *Atlantis* found 56°C. In 1967 David Ross and Egon Degens cored the sediments of the Red Sea with the research vessel *Chain* and found that the sediments were rich in oxides of copper, iron, manganese, zinc, and other metals. In 1969 Degens and Ross published their account of *The Hot Brines and Heavy Metal Deposits of the Red Sea*, an unusual paradigmic contribution to science in that it was not only scientifically revolutionary but also beautifully illustrated.[22] Interestingly, contemporary technology allowed only the oxidized brines and their metalliferous precipitates to be revealed; the sulfidic components had to await advances in deep-sea drilling technology.

Other submarine geothermal systems were described at the same time, such as Iron Ore Bay in Santorini. These were all relatively shallow water deposits, which meant that they were temperature-limited: the boiling point of water constrained the maximum temperature since, as we have seen, boiling effectively removes water and leads to the precipitation of minerals. These shallow water systems were also affected by the present highly oxygenated state of the Earth's atmosphere. This means that the sulfides, such as pyrite, tend to be oxidized in these systems.

Deep-Ocean Hydrothermal Vents

Geothermal springs in the deep ocean are better called *hydrothermal vents*. This is because the great pressures at the ocean floors, caused by the huge weight of seawater, prevent water from boiling. Just as you climb a mountain, the temperature of the boiling point of water decreases because the pressure decreases; the opposite effect occurs in the deep oceans. Indeed, the pressure at more than 1,000 m depth means that water may never actually boil. Boiling marks the change of state from a liquid to a gas and, at the boiling point, the two states coexist. Water at high temperatures may reach a critical point if the pressure is high enough to stop steam from being produced. Above this critical point, increased temperature simply causes water to become supercritical and exist as a single-phase fluid with properties between a gas and a liquid. The critical point for pure water is 374°C and 218 atmospheres pressure (about 22 megapascals in modern official units). However, the salts dissolved in seawater raises the critical point to around 407°C and 298 atmospheres. The important precept here is, as we pointed out in Aristotelian terms earlier, that boiling effectively removes water, leading to the precipitation of solids like pyrite. If boiling is prevented, water is not lost and the solids are not precipitated. This means that seawater can circulate in the ocean crust at temperatures up to around 390°C at 2.5 km depth.

In the oceans, the critical pressure of 298 atmospheres is equivalent to 2940 m water depth. In fact, until recently this depth has been at the limits of the range of the deep-sea submersibles, so the deep-sea hydrothermal vents observed have been subcritical fluids. With improved technology and the use of remotely operated vehicles, it has been possible to access hydrothermal fields deeper than 2940 m and observe supercritical water debouching from the vents. These fluids were first sampled from around 3 km depth at 5°S on the Mid-Atlantic Ridge by Andrea Koschinsky-Fritsche and her colleagues in 2008.[23]

There had been many early indications of deep-ocean hydrothermal vents before they were discovered in 1979. Indeed, it is possible that Johannes Joachim Becher sketched deep-ocean volcanic vents in 1719 (Figure 5.8). Undersea volcanoes were also well known. I am old enough to remember the emergence of the volcanic island of Surtsey, part of the Vestmannaeyrar (Vestmann Islands), from the sea south of Iceland in 1963. I visited Heimaey in the Vestmann Islands during the famous Eldfell eruption in 1973.

The discovery of hydrothermal vents in the deep ocean was made possible by improvements in submersible technology. The average depth of the ocean floor is a tad over 4 km. The only vessels able to transport people to these depths were the bathyspheres, basically metal spheres that could withstand the huge pressures. These were limited in maneuverability and

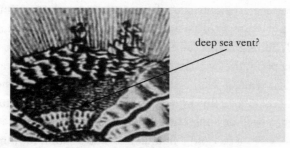

deep sea vent?

FIGURE 5.8. The first deep-sea vent prediction in 1719(?) Detail of Johannes Joachim Becher's world (Figure 5.4) showing a hot, volcanic emanation into the bottom of the ocean. Becher had drawn this at the center of the oceanic rocks, rather predicting the midoceanic rifts that separate lithospheric plate boundaries. This prescient sketch even shows marine research vessels . . .

sampling capabilities. They generally went straight down, had a look, and went back up again. In the late 1950s a series of submersibles were made, the most famous of which is the *Alvin* (Figure 5.9). The *Alvin*, which is still in operation, is essentially a titanium sphere with decking that can be driven like a submarine. It was designed and deployed by the US Navy to recover lost atomic weapons and nuclear submarines from the ocean depths. It has a series of mechanical arms that can grab things off the sea-floor and bring them up to the surface in baskets strung along the front of the vessel.

In 1977 an expedition from the Woods Hole Oceanographic Institution led by Richard Von Herzen and Robert Ballard discovered low-temperature hydrothermal vents near the Galapagos Islands in the Eastern Pacific.[24] The

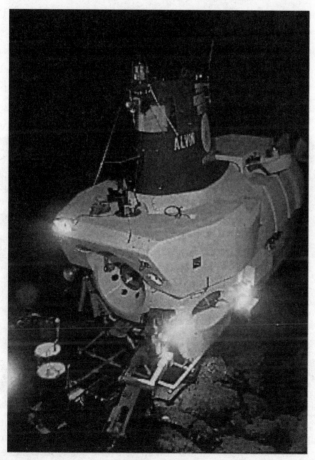

FIGURE 5.9. The submersible *Alvin* at work on the ocean floor. Photo courtesy HOV Alvin © Woods Hole Oceanographic Institution.

vents were just 8°C, but this is far higher than the normal, near-freezing 2°C of deep-ocean water. More significantly, these vents were surrounded by giant clams up to 30 cm in size (Figure 5.10).[25] This was unusual in the near-barren wastes of the abyssal plain and the Woods Hole team, true to their New England provenance and US capitalist genes, thought they could corner the market for traditional clambakes. In fact, when they opened the clams on board ship, they found they were full of hydrogen sulfide. John Edmonds, the cruise geochemist, realized that seawater had to be mixing with another fluid at these vents. He calculated that this would be a sulfide-rich fluid at 350°C.[26]

In 1979 *Alvin* pilot Dudley Foster drove scientists Bill Nordmark and Thierry Juteau along the East Pacific Rise at 21°N. They spotted a spire of rock about 2 m tall sticking out of the seafloor. A thick jet of black fluid spewed out of the top like smoke from a chimney according to Foster (Figure 5.11). The unsung heroes of 20th-century ocean science were the *Alvin* pilots. It is appropriate that one of them, Dudley Foster, is the original source of the name *black smoker*.

Foster stuck a temperature probe into the black smoke using the *Alvin* manipulators. It showed 32.7°C, the maximum temperature it

FIGURE 5.10. Black smokers are oases of life in the otherwise desolate abyssal plains of the deep oceans. The biomass around black smokers is comparable with that in the Amazon jungle. Over 500 new species have been found since their discovery in 1979, including these giant clams and the sea anemone. Photo courtesy HOV Alvin © Woods Hole Oceanographic Institution.

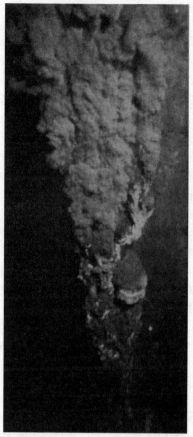

FIGURE 5.11. A 2-m black smoker from the East Pacific Rise at 21°N. Photo courtesy HOV Alvin ©Woods Hole Oceanographic Institution.

could measure. Foster assumed the probe was not working and, when *Alvin* returned to the surface, *Alvin*'s engineer Jim Akens found that the plastic tip of the probe had melted. This plastic melted at 180°C and was the same plastic as in the *Alvin*'s windows. Foster and his scientists had been very lucky. Akens made a new probe and 350°C was measured in the black smoke on the next dive. This was the temperature predicted by John Edmonds.

Since that discovery, black smokers have been found in all the Earth's oceans. They are particularly abundant along the midocean ridges, a giant 60,000-km-long mountain range that girdles the Earth rather like the stitching on a baseball. These ridges delineate zones where the geotectonic plates that constitute the Earth's outer skin are splitting apart. New crust is being made by extensive volcanic activity and intrusion of molten rock from the underlying mantle. Black smokers are also found at the other side of geotectonic plates where they are being destroyed and are sliding back down into the underlying mantle of the Earth. Again, these zones are extremely volcanically active but, by contrast with the midocean ridges, they are characterized by deep trenches and, since they are nearer the continents, thick sediments as well as volcanic rocks cover them.

The *Alvin*, which did most of the original exploring, worked comfortably to a depth of 2,000 m in the oceans. Below this, the pressure was too high for extended periods of exploration to be safely achieved. Unfortunately, the mean depth of the ocean floor is some 4,000 m, so most of the oceans were outside the original range of the submersible. This is one reason why the early accounts of black smokers were mainly limited to the midocean ridges, the chain of enormous undersea mountains that rise to over 2,000 m and, in some places, above the sea surface, forming oceanic islands. The new generation of deep-sea submersibles, including the reequipped *Alvin*, are able to survey these greater depths. In addition, the deployment of sophisticated remotely operated vehicles has allowed extreme depths to be explored. In 2013, a UK team discovered black smokers at 4,968 m depth in the Cayman trough. This is the deepest black smoker yet discovered and this was achieved with a remotely operated vehicle.

The immediate importance of these deep-sea hydrothermal vents is that they show how the Earth loses its heat. Heat is generated in the Earth mainly by radioactive decay. The deep-ocean hydrothermal vents are fed by ocean water penetrating the ocean crust and carrying the heat away. The analogy is the water-cooled automobile engine. The seawater is maintained at a temperature just beneath the critical point where the conductivity is at a maximum. The deep Earth heat is carried to the upper parts of the crust by volcanic activity. These hot rocks warm the cold, circulating seawater and carry the heat into the oceans. Ultimately it is radiated into space by the ocean–atmosphere system.

The black color of the hydrothermal vent fluids is caused by their content of black iron sulfide particles, including pyrite. The hot fluids hit the ice-cold ocean water and the sulfides precipitate. The heat produced at the vent point sources is so large that it powers convection in the oceans, leading to huge circulation patterns or gyres. In fact, these systems were predicted by Kurt Boström and Melvin Peterson in 1966 when they described anomalous concentrations of manganese across a huge area of the equatorial Pacific.[27] The paper was initially rejected because the area they found was asymmetric. In fact it turned out it was the result of the high heat flow of the hydrothermal vents and the rotation of the Earth.

In 2009 George Luther and his colleagues from the University of Delaware showed that the black smoke contained nanoparticulate pyrite,[28] like the pyrite dust shown in Figure 4.16 in Chapter 4. These particles are so small that they do not sediment in the oceans. Rather they are dispersed into the ocean water and oxidize gradually to release iron. The modern oceans are highly deficient in iron, and the concentration of dissolved iron limits the biologic productivity of the oceans. The pyrite in the black smoke from the hydrothermal vents provides an important source of this micronutrient to the ocean and helps drive the Earth's ecology and environment. We mentioned in Chapter 2 that pyrite fed the world through its role in providing the sulfur for fertilizer production. Here we can see that pyrite feeds the ocean too, releasing biologically available Fe for enhanced primary biological productivity.

Black Smoker Chimneys and Massive Pyrite Deposits

The hot waters debouching from the deep-ocean vents are black with iron sulfides, including pyrite, and the points from which they exit are large, chimney-like structures often mainly consisting of pyrite. These chimneys can build up to large edifices on the seafloor. The tallest yet observed grew to more than 40 m in height[29] at rates of up to 9 m in eighteen months. More commonly they are smaller, perhaps 1 to 2 m in height, and occur in groups.

Pyrite-rich chimneys were found before the high-temperature hydrothermal vents. They were discovered in 1978 on the East Pacific Rise at 21°N by a French–American–Mexican team led by Jean Francheteau using the French submersible *Cyana* operated by IFREMER, the French Oceanographic Institute. The group of chimneys found was extinct: that is, the chimneys had no hydrothermal fluid coming out of them. However, Francheteau and his colleagues realized that these chimneys were precursors to sulfide ore deposits.

The connection between these chimneys, hydrothermal vents, and sulfide ore deposits was established in the famous 1982 NATO conference in Cambridge, UK (Figure 5.12). Here all the pioneering figures in the deep-ocean vent sulfide business were present.[30]

The chimneys form mainly from the precipitation and crystallization of pyrite when the hot hydrothermal fluid hits cold seawater (Figure 5.13). The pyrite crystals themselves do not form robust piles sufficient to form these tall edifices. They are cemented by calcium sulfate, which is a major component of the everyday cement used in the building trade. The unusual

FIGURE 5.12. Participants in the NATO Advanced Research Institute. Hydrothermal Processes at Seafloor Spreading Centers, April 5–8, 1982, at Cambridge University, England (see endnote 30).

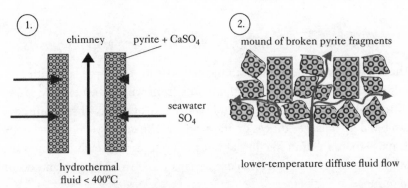

FIGURE 5.13. Construction and collapse of black smoker chimneys. Calcium sulfate ($CaSO_4$) precipitates between the pyrite crystals as the seawater sulfate (SO_4) is entrained in the hydrothermal fluid. As the hydrothermal system cools, the calcium sulfate dissolves and the chimney collapses to form a mound of pyrite rubble. Lower-temperature hydrothermal fluids continue to flow through the pyrite rubble, indurating it and forming a massive sulfide.

thing about calcium sulfate is that it gets less soluble as it gets hotter. This is the reverse of usual solubility: if you want to dissolve something you generally heat it or mix it with hot water. So as the hot hydrothermal fluid rises it entrains seawater, which contains calcium and sulfate. This seawater heats up and calcium sulfate precipitates, cementing the pyrite and forming these large, chimney-like edifices above the vent.

The whole point of the hydrothermal vent is to take heat away from the hot volcanic rocks below. After a time these rocks cool and solidify and the heat source is removed. So the chimney cools down. Since calcium sulfate has this odd reverse solubility, it begins to dissolve. As the cement dissolves the chimney tumbles, forming a pile of pyrite rubble. The hydrothermal system does not entirely switch off, and lower-temperature fluids circulate in the interstices between the pyrite rubble blocks, dissolving old pyrite and precipitating new. This continues until an indurated mass of mainly pyritic sulfides is produced. It is this indurated mass of pyrite-rich rubble that we recognize as massive sulfide deposits in ancient rocks.

Massive Pyrite Deposits as Fossil Hydrothermal Systems

Veins are potentially rich but small-scale mining objects. With increased industrialization in the 19th century, the need for large-scale metal production became paramount. Advances in mining technology and metallurgy allowed the large but lower-grade, massive pyrite deposits to be exploited. The origin of many of these deposits became contentious. In particular, they often did not cross-cut the country rock like veins but appeared to be bedded with sediments and volcanic products like tuffs and lavas (Figure 5.14).

The massive pyrite deposits were thought to have been formed at the same time as the enclosing rocks and were called *syngenetic*. This contrasts with veins that are obviously formed later than the country rock and are called *epigenetic*. A school developed in France after the Second World War that took syngenesis as a dogma and fitted all sulfide ore deposits to this. Sir Kingsley Dunham, the great director of the British Geological Survey, remarked that this was the reasoning of the Schoolmen. By contrast, a substantial school of thought especially based in the United States continued to promote the idea that all these deposits were epigenetic—it was just that some were intruded along the bedding planes of sediments and volcanics and appeared to be syngenetic.

The syngenetic idea was extremely important to exploration geologists and mining engineers because massive sulfides with a bed-like form contained far more metal than any cross-cutting vein. For example, an average stratiform massive sulfide might contain 10 million tons of ore and some of the great deposits well over 100 million tons. This compares with the less than 1 million tons for the average vein. The relative ease of extraction was also an issue. A stratiform deposit can be mined using horizontal techniques, allowing large mining machines access to the face and relative

FIGURE 5.14. Pyrite-rich sediments from the c. 20-million-year-old Kuroko district, Japan. The pyrite is forming classical sedimentary structures and shows graded bedding with coarser grains toward the base. The 20th-century debate was whether these textures showed that pyrite was formed in volcanic sediments or hot, pyrite-bearing solutions had penetrated the rocks later, mimicking the earlier sedimentary structures (see color plate). Photo William Sacco courtesy of the Society of Economic Geologists. From H. Ohmoto and B.J. Skinner, eds., 1983. The Kuroko and related volcanogenic massive sulfide deposits. *Economic Geology Monograph,* 5: cover image.

ease of transport. This compares with the smaller technology necessarily employed in the extraction of smaller and often steeply dipping veins.

In a remarkable example of scientific prescience, Chris (Tofte) Oftedahl, a Norwegian geologist who had been studying the 400-million-year-old massive pyrite deposits of the Norwegian mountains, proposed that these massive sulfide deposits were formed syngenetically together with sediments as a result of volcanism on the deep-ocean floor.[31] Together with my colleague Hans Zweifel, I extended these ideas in 1975 to the great Precambrian (i.e., formed during the first 4,000 million years of Earth history) massive sulfide ores[32] that provide much if not most of the volcano-related massive pyrite mineralization today. These conclusions were based on geologic studies of these ancient ores and were arrived at some years before the hydrothermal vents and their black smokers were discovered in the modern oceans.

Many of the massive pyrite deposits are contained in ancient volcanic rocks or are closely associated with contemporary ancient volcanism. The massive pyrite lenses are often crossed by later veins and were originally underlain by a network of veins, often copper-rich, which were conduits for the hot fluids that formed the massive pyrite. This is the reason for the controversy around these deposits as to whether they were syngenetic or epigenetic. They clearly display epigenetic characteristics since the veins are cross-cutting the earlier host rocks. However, what we originally proposed was that the veins and ore lenses were part of the same system on the seafloor and not a result of a separate event many millions of years later. Earlier it had been thought that the veins and ore lenses had replaced the host rock and were formed by hot metal-rich fluids emanating from later granitic intrusions into the country rocks.

I remember being interrogated about this idea by one of the great ore geologists of the day, Sven Gavelin. Sven, who was around sixty-five at the time and could not understand how this callow youth could argue that the massive sulfides were syngenetic when it was clear that many sulfide lenses cross-cut the host rocks and were obviously later. To him it was like arguing that black was white. I had a beautiful specimen of folded massive sulfide on the desk between us that showed that the massive sulfide had been emplaced before the folding of the rocks occurred. That is, the ore had been formed contemporaneously with the formation of the country rocks. The cross-cutting features were all part of the hydrothermal system and not evidence for some separate, later event. The weakness in my argument was that black smokers had not yet been discovered in the modern oceans. This was to be rectified some four years later.

Although the geologic setting, mineralogy, chemistry, and form of massive sulfides are indicative of deep-ocean hydrothermal vent processes, other key indicators have been described that prove this. These include ancient chimney fragments and fossilized vent fauna, as well as isotopic fingerprints. This means that each massive pyrite deposit is a fossilized deep-ocean hydrothermal vent system. These massive pyrite deposits have been found throughout the geologic record. They appear in the oldest well-preserved volcanic settings, such as the 3.5-billion-year-old rocks of the Pilbara region of northwest Australia. These massive pyrite deposits suggest that the same oceanic circulation processes were cooling the Earth during most, if not all, of geologic history. These ancient black smoker relicts provide a window on the ancient Earth and its geochemical and biological environments.

If the massive pyrite deposits are related to volcanism, then the distribution of ancient massive pyrite ores is related to the distribution of ancient volcanoes. It became apparent during the plate tectonic revolution in the Earth sciences that the distribution of volcanoes was closely related to geotectonic processes. This had two consequences. From the point of view of exploration for new ore deposits, the detailed mapping of ancient volcanic systems provided good guides to massive sulfide deposits. On the other hand, massive pyrite ore deposits were themselves easily recognizable and could provide information about the volcanic and therefore the tectonic environment of the period. Volcanic-related massive pyrite deposits thus became major players in understanding plate tectonics and, especially, the ancient history of plate tectonics.

Notes

1. G. Agricola. 1556. *De re metallica*. Translated by H. Hoover and L. Hoover, 1950 (New York: Dover), p. 46.
2. Dryness and dampness were active properties whereas heat and cold were passive. These properties existed in binary active/passive combinations.
3. E.R. Caley and J.F.C. Richards. 1956. *Theophrastus on Stones*. Graduate School Monographs (Columbus: Ohio State University Press), 238pp.
4. *Nützlich Bergbüchlin* published anonymously in 1527, printed by Johan Loersfelt in just 24 pages. Agricola refers to a certain Calbus Freibergius at the end of Book III of *De re metallica*. His statement of Calbus's views are astonishingly similar to statements in the *Nützlich Bergbüchlin* and leave little doubt that this Calbus was the author of that anonymous book on veins. Calbus Freibergius is almost certainly the Freiberg Dr. Rühlein von Kalbe; he was a doctor and burgomaster at Freiberg at the end of the 15th and the beginning of the 16th centuries. He was also a mathematician who helped to survey and design the mining towns of Annaberg in 1497 and Marienberg in 1521.
5. M. de Maillet. 1721. *Telliamed*. Translated by A.V. Carozzi, 1968 (Urbana: University of Illinois Press). *Telliamed* is, of course, *de Maillet* spelled backward. This was an extremely popular book published some ten years after de Maillet's death but widely circulated in manuscript form before this.
6. J.F. Henckel. 1725. *Pyritologia, oder: Kieß-Historie, als des vornehmsten Minerals, nach dessen Nahmen, Arten, Lagerstätten, Ursprung* (Leipzig: J. Chr. Martin), p. 81.

7. Herodotus. *The Histories*. Book 2. Translated by R. Waterfield, 2008 (Oxford: Oxford University Press), 840pp. The ideas regarding the age of the Earth widely attributed to Herodotus are actually much more vaguely represented in the actual text. Herodotus is in the middle of an argument about defining the land of Egypt when he mentions this calculation in passing. He is arguing against an idea that Egypt historically is merely the Nile delta.

8. *Meteorologica*. Translated by E.W. Webster, 1931. In W.D. Ross, ed. *The Works of Aristotle* (Oxford: Clarendon Press), pp. 21–195.

9. Most students are taught that Steno invented the Law of Superposition in 1669, but Avicenna wrote in the *Kitab Al-Shifa* in 1027:

    > It is also possible that the sea may have happened to flow little by little over the land consisting of both plain and mountain, and then have ebbed away from it.... It is possible that each time the land was exposed by the ebbing of the sea a layer was left, since we see that some mountains appear to have been piled up layer by layer, and it is therefore likely that the clay from which they were formed was itself at one time arranged in layers. One layer was formed first, then at a different period, a further was formed and piled, upon the first, and so on. Over each layer there spread a substance of different material, which formed a partition between it and the next layer; but when petrification took place something occurred to the partition which caused it to break up and disintegrate from between the layers (possibly referring to unconformity)....
    > As to the beginning of the sea, its clay is either sedimentary or primeval, the latter not being sedimentary. It is probable that the sedimentary clay was formed by the disintegration of the strata of mountains. Such is the formation of mountains.

10. Athanasius Kircher. 1664. *Mundus Subterraneous*. 2 vols. (Amsterdam: L. Figuier). The average increase in temperature with depth near the Earth's surface is around 25°C per kilometer.

11. None of Thales's writings have been preserved, but he has been called the first scientist.

12. I. Newton. 1704. *Opticks or a Treatise of the Reflections, Refractions, Inflections and Colours of Light* (London: William Innys), 382pp.

13. N. Lémery. 1700. Expliquation physique et chymique des feux souterrains, des tremblements de terre, des ouroganes, des éclairs et du tonnere. *Mémoires de l'academie royaume, Paris*, 101–110.

14. D. Rickard and R.O. Hallberg. 1973. De små gruvarbetarna i Falun. *Svensk Naturvetenskap*, 102–107.

15. R. Descartes. 1991. *Principia Philosophiae: Part 4. The Earth*. Translated by V.R. Miller and R.P. Miller (Dordrecht: Kluwer Academic Publishers).

16. J. Bostock and H.T. Riley. 1855. *Pliny the Elder, The Natural History*, Vol. 31 (London: Taylor and Francis).

17. For example, F.W. Von Oppel. 1749. *Anleitung zur Markscheidekunst* (Dresden: GC Walther).

18. R.H. Sibson, R.M. Moore, and A.H. Rankin. 1975. Seismic pumping—a hydrothermal transport mechanism. *Journal of the Geological Society, London*, 131:653–659. J.G. Ramsay. 1980. The crack-seal mechanism of rock deformation. *Nature*, 284:135–139.

19. P. Torssander. 1986. *Origin of Volcanic Sulfur in Iceland: A Sulfur Isotope Study* (PhD diss., Stockholm University).

20. R. Bunsen. 1847. Ueber den innern Zusammenhang der pseudovulkanischen Erscheinungen Islands. *Annalen der Chemie und Pharmacie*, 63:1–59.

21. See, for example, D. Rickard and G.W. Luther. 2007. Chemistry of iron sulfides. *Chemical Reviews*, 107:514–562.

22. E.T. Degens and D.A. Ross, eds. 1969. *Hot Brines and Heavy Metal Deposits in the Red Sea: A Geochemical and Geophysical Account* (Heidelberg: Springer-Verlag), 600pp.

23. A. Koschinsky, D. Garbe-Schönberg, S. Sander, K. Schmidt, H.H. Gennerich, and H. Strauß. 2008. Hydrothermal venting at pressure-temperature conditions above the critical point of seawater, 5°S on the Mid-Atlantic Ridge. *Geology*, 36:615–618.

24. J.B. Corliss, J. Dymond, J.M. Edmonds, et al. 1979. Submarine thermal springs on the Galapagos Rift. *Science*, 203:1073–1083.

25. P. Lonsdale. 1977. Clustering of suspension-feeding macrobenthos near abyssal hydrothermal vents at oceanic spreading systems. *Deep Sea Research*, 24:857–863.

26. J.M. Edmonds, C. Measures, R.E. McDuff, et al. 1979. Ridgecrest hydrothermal activity and the balances of the major and minor elements in the ocean: the Galapagos data. *Earth and Planetary Science Letters*, 46:1–18.

27. K. Boström and M.N.A. Peterson. 1966. Precipitates from hydrothermal exhalations on the East Pacific Rise. *Economic Geology*, 61:1258–1265.

28. M. Yücel, A. Gartman, C.S. Chan, and G.W. Luther. 2011. Hydrothermal vents as a kinetically stable pyrite (FeS_2) nanoparticle source to the ocean, *Nature Geoscience*, 4:367–371.

29. Discovered by a University of Washington team in 2011 on the Axial Seamount in the Pacific Ocean at 1,900 m depth. Godzilla, the great >45-m-tall black smoker in the Pacific Ocean off Oregon, was discovered in 1991 and collapsed in 1995.

30. Front row (left to right): H. Craig, D.S. Cronan, J. Francheteau, C.R.B. Lister, G. Thompson, K.C. Macdonald, F. Machada, P.A. Rona, J. Honnorez, R.F. Dill, R.D. Ballard, N.A. Ostenso, R. Hessler, H. Thiel, and F. Grassle. Second row (left to right): J. Verhoef, R Whitmarch, V. Stefansson, B.E. Parsons, T. Juteau, G.A. Gross, H.P Taylor Jr., F. Albarede, H. Jannasch, E. Bonatti, K. Crane, J. Lydon, I.D. MacGregor, and E.R. Oxburgh. Third row (left to right): R. Hekinian. B.J.

Skinner, C. Mevel, L. Widenfalk, R. Bowen, H. Bougaull, T.H. van Andel, J.R. Cann, R.J. Rosenbauer, D. Rickard, A. Malahoff, S.P. Varavas, and M.J. Mottl. Fourth row (left to right): K. Brooks, J.W. Elder, B. Stuart, K. Gunnesch, A. Fleet, H.T. Papunen, A.H.F Robertson, S.A. Moorby, J. Boyle, C. Lalou, and V. lttekott. Top row (left to right): K.K. Turekian, J. Hertogen, J.A. Pearce, J.M. Edmond, S.D. Scott, D.B. Duane, A.S. Laughton, H.-W. Hubberton, R. Chesselet, and R.L. Chase.

31. C. Oftedahl. 1958. A theory of exhalative-sedimentary ores. *Geologiska Föreningens I Stockholm Förhandlingar*, 80:1–19.

32. D. Rickard and H. Zweifel. 1975. Genesis of Precambrian sulfide ores, Skellefte District, Sweden. *Economic Geology*, 70:255–274.

6

Microbes and Minerals

PYRITE CONSISTS OF two elements—iron and sulfur—but considerations of pyrite formation have mainly concerned sulfur. Iron is extremely abundant in the Earth; in fact, it is the fourth most abundant element on Earth and is less localized in its distribution. By contrast, sulfur is the 17th most abundant element in the Earth's crust, and there are about 100 times more iron than sulfur. The interest in the primary role of sulfur in pyrite formation continues to the present day.

In the Old Latin of early Republican Rome (i.e., before c. 200 BCE), sulfur was called *sulpur* or burning stone (i.e., brimstone). The *p* was pronounced with a puff of air. This puff was transliterated with an *h* following the *p*. When the *f* sound was introduced into classical Latin, *p* was often changed to *ph* in Latin words of Greek origin. *Sulpur*, however, had no Greek roots.[1] The Greeks called it θεῖον (*thion*), which gave rise to our prefix, *thio-*. *Sulfur* had been written as *sulphur* in Old Latin, with the *h* indicating the puff of air after the *p*, but when the *f* sound was introduced this gave the mistaken impression that *sulphur* was originally a Greek word. At the end of classical times (around 27 BCE) the spelling was altered to *sulfur*, which is the spelling that usually appears in Latin dictionaries. In the last millennium, the element has traditionally been spelled *sulphur* in the United Kingdom and countries where UK rule held sway. By contrast, US English has continually used the correct *sulfur* spelling. The fountainhead of all chemical definitions worldwide, the International Union of Pure and Applied Chemistry, adopted the spelling *sulfur* in 1990. Finally, the UK authorities admitted their error and the UK Royal Society of Chemistry Nomenclature Committee recommended the correct spelling in 1992. In 2000 the authority determining quality and standards in UK schools decreed UK children should be taught the *sulfur* spelling. The *sulphur* spelling still occurs, but at best this is a literary affectation.

We have seen that the ancients observed that sulfur originated in volcanoes and their associated hydrothermal features such as geothermal springs and deep-ocean vents. They had also observed that pyrite formed, especially as veins, in sedimentary rocks. In these settings, the origin of sulfur led to a series of wild speculations concerning the interior of the Earth and the nature of minerals. Pyrite was classically regarded as a partially evolved substance that was on its way to becoming gold. It was formed initially in the center of the Earth and then found its way to the surface. Many supposed that the veins grew in the Earth. The longer the journey, the more it evolved and became richer in gold content.

Rotten Eggs

Even so, there was awareness in the ancient world that sulfur could occur in nonvolcanic environments. For example, Pliny described the poisonous sulfurous vapors that could be encountered in water wells. Agricola described these poisonous sulfurous vapors occurring in some mines:

> There are also times when a reckoning has to be made with Orcus [i.e., Pluto, the god of the underworld, my explanation] for some metalliferous localities, though such are rare, spontaneously produce poison and exhale pestilential vapor, as is also the case with some openings in the ore, though these more often contain the noxious fumes. In the towns of the plains of Bohemia there are some caverns which, at certain seasons of the year, emit pungent vapors which put out lights and kill the miners if they linger too long in them. Pliny, too, has left a record that when wells are sunk, the sulfurous or aluminous vapors which arise kill the well-diggers, and it is a test of this danger if a burning lamp which has been let down is extinguished. In such cases a second well is dug to the right or left, as an airshaft, which draws off these noxious vapors. On the plains they construct bellows which draw up these noxious vapors and remedy this evil; these I have described before.

The poisonous and foul-smelling gas was hydrogen sulfide, which reacts with iron compounds to form pyrite. Elemental sulfur reacts with iron to form pyrite only at elevated temperatures. At lower temperatures the reaction is slow. By contrast, the reaction between iron salts and hydrogen sulfide to form pyrite occurs in water and is relatively fast. So hydrogen sulfide is the main source of the sulfur in pyrite.

Hydrogen sulfide was known to the ancients as sulfurous vapor or divine gas. It may be that the ancient Greek word for sulfur (transliterated as *theion*) also means "divine."[2] It was first identified as a distinct chemical compound by the Swedish chemist Carl Wilhelm Scheele in 1777. Scheele produced the gas in the laboratory by the reaction of iron sulfide, the mineral pyrrhotite, with mineral acids. Older readers will be familiar with this process from their school chemistry labs. Hydrogen sulfide was a key reactant in a series of experiments aimed at discovering the composition of unknown substances, in the days before machine-based analysis. Unlike today, you could not buy ready-made hydrogen sulfide gas in convenient gas bottles. You produced it in Kipp's apparatus: basically a glass reaction vessel where acid was poured onto granules of the iron sulfide, pyrrhotite. By contrast with pyrite, which is inert to mineral acids, pyrrhotite rapidly dissolves and produces hydrogen sulfide.

Scheele has been called "hard-luck Scheele" since he discovered oxygen, molybdenum, barium, hydrogen, and chlorine before people like Humphry Davy, Joseph Priestley, and Antoine Lavoisier who are generally given credit today. He called hydrogen sulfide *Schwefelluft* or sulfur air and described it as *stinkende* (smelly). In 1778, his patron Torbern Olaf Bergman demonstrated that hydrogen sulfide (which he called *hepatic air*) occurred in medicinal mineral springs around Orebro in middle Sweden.[3] In 1819 Jons Jacob Berzelius[4] discovered that hydrogen sulfide was the sulfur equivalent of water. It has the formula H_2S, compared with H_2O for water.

The definition of H_2S coincided with an increasing concern about deaths in the sewers of Paris. In 1806 these deaths were related to the presence of H_2S by Baron Guillaume Dupuytren, a French military surgeon better known for treating Napoleon Bonaparte's hemorrhoids. H_2S is exceptionally poisonous. For example, it is about ten times more poisonous than hydrogen cyanide, that popular killer in murder mysteries. The reason why H_2S has not provided a popular toxin in detective fiction is because the human nose can detect the smell of H_2S at concentrations as low as 0.5 parts of H_2S in 1 billion parts of air. You can detect one molecule of H_2S in a normal-sized drawing room or, to continue the detective fiction analogy, in the library.[5] Normally, 10 parts of H_2S in 1 million parts of air is accepted as the level at which prolonged exposure becomes toxic.

Eight hundred parts of H_2S in 1 million parts of air causes rapid death, although lower concentrations can be lethal. These are between 4 and 10 times lower concentrations than for similar effects from hydrogen cyanide.

I have worked with H_2S for many years and so have had the opportunity to study the effects of H_2S poisoning first hand. The first thing you notice is the smell. In low concentrations, H_2S smells like rotten eggs. As the concentration increases, the smell changes. It becomes sweeter and rather sickly. Interestingly, at higher concentrations you cannot smell it at all. This is a symptom I recommend you take particular notice of since, shortly after that happens, you drop down dead. I kept an oxygen cylinder at hand in the lab for emergency use in this case. Breathing pure oxygen oxidizes the H_2S.

I thought that H_2S reacts with the metal proteins, such as hemoglobin, in the blood and interferes with oxygen transport. This had originally been proposed by the great German biochemist Felix Hoppe-Seyler in 1863.[6] I can see now that suffocation would not accord with the suddenness of my demise after prolonged exposure to H_2S. In fact, it has become apparent that the causes of H_2S toxicity in humans and other organisms are complicated and still not entirely understood. Rather the effects seem to be mainly targeted on the enzyme cytochrome oxidase. Additional effects are due to sulfide attacks on brainstem respiratory nuclei and irritation of the lungs leading to pulmonary edema.[7]

Hydrogen sulfide is particularly common in limestones, and many deaths still occur from H_2S poisoning in wells in limestone terrains. Limestones are good aquifers, and farmers needing water can sink wells fairly easily into the relatively soft rock. Usually what happens is the man digging at the bottom of the well is overcome by H_2S poisoning. His colleague goes down to save him and is also overcome. These H_2S-bearing limestones are often called stinkstones and are featured in geologic descriptions and texts from the 18th century. However, it seems that the New Zealand chemist William Skey was the first to prove that the odor emitted when stinkstone is struck is caused by hydrogen sulfide.[8] Prior to that report in 1892, the smell was thought to be due to organic compounds. At least part of the sulfide contained in limestones may be original sulfide trapped in the sediments. However, much of the free hydrogen sulfide in sedimentary rocks is associated with hydrocarbons (e.g., sour gas). Interestingly, it took some time for the scientific community to relate these H_2S-bearing limestones with the formation of sulfide veins. It seems obvious now that metal-rich groundwaters traveling along fissures in the limestone react with this hydrogen sulfide to precipitate metal sulfides. In fact, it is probable that many of the lead and zinc sulfide veins that occur in limestones were formed in this way. However, in 1979 I could still publish a research paper on this process as an original idea.[9]

Blue Mud

Because of its obvious smell and our built-in nasal super-detector, the occurrence of hydrogen sulfide in modern sediments must have been well known before it was scientifically documented. However, as detailed later, the scientific community had to wait until the results of the *Challenger* expedition to be published in 1891 for the abundance of H_2S in modern sediments to be documented. I always found this surprising, but, even today, few people realize how abundant hydrogen sulfide is in sediments. Go down to a local ditch, pond, or stream, or even a seashore or estuary, with a bottle of vinegar or mineral acid. Dig into the black mud and add the vinegar. You will soon smell bad eggs, characteristic of hydrogen sulfide. The H_2S is released from the pore waters in the muds by the acid in the vinegar.

Natural processes may also release H_2S from the muds. These are usually catastrophic events because of the toxicity of H_2S. Local, usually storm-induced, sediment overturns in these zones can lead to disastrous results. For example, I kept a cutting in my wallet for many years from the now-defunct UK national newspaper *The News Chronicle*. In 1960 the reporter described such an event off Walvis Bay, Namibia. They described how the town hall clock in Swakopmund turned black due to H_2S and "sharks came up gasping on the evening tide."

During your expedition to the local mud you may find a layer of bluish-gray–colored mud usually at a greater depth than the black layer. These blue muds are easily distinguished from the more abundant black and brown muds. In 2005 I showed, together with my colleague John Morse, that the bluish-gray color of these muds is correlated with the presence of finely divided pyrite.[10]

Blue muds are abundant in the oceans and were used as indicators of position before global positioning satellites and accurate clocks. The old-time sailing masters not only kept sketches of coasts and ports they had visited but also notes on the color of the seawater and the nature of the sediments. The character of the seafloor was regularly sampled, usually by means of tallow on the end of a weighted rope, recorded and used to help determine position. Blue mud was commonly reported as a fine-grained muddy deposit with a light, distinct bluish-grey color.

In 1773 two bomb ketches, HMS *Racehorse* and HMS *Carcass,* were dispatched on a scientific voyage to explore the Arctic Ocean. Their specific task was to sample the seafloor and report the sediments in order to provide better guidance to mariners in inshore waters. Captain the Honourable John Constantine Phipps was in overall command. Fourteen-year-old Mr. Horatio Nelson was midshipman on the *Carcass,* whose captain was Skeffington Lutwidge, later Admiral of the Red. Nelson was famously reprimanded by Skeffington Lutwidge for rashly hunting polar bears instead of attending to his duty. He had desperately wanted to take a polar bear skin home to his father. Unfortunately his musket misfired and Nelson set off after the bear, intending to bludgeon it with the musket stock. Luckily the *Carcass* fired a shot that scared the bear off. The expedition weighed anchor on June 4, 1773. By July 31, 1773, they were stuck fast in the ice northeast of Spitsbergen and were back in port by September 17, 1773. The expedition took soundings of the ocean floor on the continental shelf in the Norwegian Sea, possibly the first such soundings recorded from such depths. The Master reported blue mud from 383 fathoms.

Almost exactly 100 years later, the Royal Society of London equipped a Royal Navy ship, HMS *Challenger,* for a scientific expedition. The purpose of this 1872 expedition was to circumnavigate the globe surveying and sampling the oceans and the seafloor sediments. The expedition took four years and traveled 130,000 km (80,000 miles). In doing so it laid the foundations of oceanography. The publication of the results of the voyage took many years to complete and catalogued 4,000 new species, among other things.

John Murray was originally appointed as an assistant to the expedition's principal scientist Charles Wyville-Thomson. When Wyville-Thomson retired in 1879 on health grounds, Murray took over the job of publishing the results of the expedition. This he completed in 1896. For this reason Sir John Murray has always been primarily associated with the *Challenger* expedition and is regarded as the "father of oceanography." However, the original brains behind the expedition was Wyville-Thomson. Wyville-Thomson was an early advocate of Charles Darwin and proposed that dredging the sea bottom might reveal modern species that had affinities with fossil ancestors. In an earlier cruise, Wyville-Thomson had discovered a calcareous ooze forming on the seafloor today that bore a striking resemblance to the 75-million-year-old Cretaceous chalk of the white cliffs of Dover.

In 1891 Murray published, together with the Jesuit priest Alphonse-Francois Renard, professor of geology and mineralogy at the University of Ghent, a ground-breaking account of seafloor samples collected by the *Challenger* expedition.[11] This was the foundation of marine geology. It is a remarkable volume. My version is a large royal quarto book bound in its original green buckram from 1891. It has 525 text pages, 29 color plates, and 19 charts. The color plates are exquisite chromolithographs of original ink and watercolor images of the rocks and animals collected in the sediment dredges. They include micrographs of manganese nodules from the sediments described in Chapter 5 in which Boström and Peterson were later to find the first indications of deep-ocean hydrothermal vents.

Murray and Renard reported H_2S commonly emanating from sounding tubes and dredges of blue muds from the seafloor. They stated that H_2S was present "in all harbor muds, muddy bays near land and, indeed, in nearly all the terrigenous deposits, such as the blue muds" and wrote that "The blue color is due to organic matter and sulphide of iron in a fine state of division."

Murray and Renard reported that sulfidic blue muds are the dominant sediment in both deep and shallow waters in all partly enclosed seas and the continental shelves and slopes. John Young Buchanan was the analytical chemist on the *Challenger* expedition and must have been responsible for most of the chemical analyses reported by Murray and his co-authors. He was the last surviving member of the expedition (d. 1925). In 1890 he reported free sulfur from twenty-seven samples of fresh, estuarine, continental shelf, and deep-ocean sediments. Buchanan disagreed with Murray and Irvine regarding the relative roles of surface oxidation versus burial in the development of the iron oxide and iron sulfide layers in many marine deposits. As an analyst Buchanan was very much aware of the risk of surface oxidation of sedimentary sulfides during sampling and considered that this was the main source of the iron oxides.[12] Murray thought that the iron oxides were originally in the sediments and resulted from different sedimentary conditions.

Source of Hydrogen Sulfide

We have seen that H_2S is common in sediments worldwide. So the next question is: Where does this H_2S come from? It is fairly obvious that it does not come from volcanic processes. On the other hand, it is not in our everyday surface environment. So it must be produced in the sediment.

Seawater contains a sulfur-bearing compound, sulfate or SO_4. Here sulfur is combined with oxygen instead of being combined with hydrogen, as is the case with H_2S. Sulfate is the third most abundant dissolved substance in seawater after sodium and chlorine: about one-tenth of the salt dissolved in normal seawater is sulfate. However, in 1895 Murray and Irving reported the surprising disappearance of sulfate from the seawater trapped in blue mud.[13] They found that sulfate was present in the seawater trapped in other sediments, but they could not find sulfate in the seawater trapped in blue muds. This was one of the first observations of the *Challenger* expedition. The *Challenger* had sailed out of the Clyde in Scotland at the beginning of its epic voyage and examined the sediments in the estuary. These were blue muds, and the trapped seawater in them was free of sulfate. They found that this was a characteristic feature of blue muds worldwide.

In 1811, William Haseldine Pepys, a distant relation of the diarist, reported that he had inadvertently synthesized pyrite when a mouse got into a laboratory jar of iron sulfate. The report appeared in the first volume of the *Transactions of the Geological Society of London*.[14] The Geological Society of London was the world's first geologic society and had evolved from the British Mineralogical Society, of which Pepys had been a founding member and erstwhile president. I wrote once that Pepys must have had an interesting chemical laboratory for mice to be running around in it. Pepys concluded that the pyrite could form by the reaction of the iron sulfate with "animal matter" and that the sulfate was "entirely deoxygenated" or *reduced* as we would say today. In simple terms, Pepys realized that, effectively, the oxygen had been removed from the sulfate. By 1838 this observation was recorded in a standard geology textbook.[15]

We noted in Chapter 4 that Agricola had reported pyrite in wood in 1556 and Henckel had described pyrite in organic matter and replacing shells in 1725. In the 19th century the pace of reports of the association of pyrite with organic matter quickened. By the time that Murray and his team were considering the sulfur chemistry of blue muds, the relationship between pyrite and organic matter was well established. It was generally thought that H_2S was one of the products of the putrefaction of organic matter.[16] In effect, organic matter rotted, H_2S was produced, and this reacted with available iron to form pyrite.

This is a very different idea from that of Pepys, however. He had suggested that the organic matter reacted with sulfate in the system to remove the oxygen and produce sulfide. The alternative idea was that the sulfur was originally a part of the organic matter itself. Sulfur is a minor component of all living matter. So, as organisms died and the organic matter decomposed, the sulfur in the organic matter would be at least partially released as hydrogen sulfide.

But you can see that this would not explain why sulfate in seawater was reduced in blue muds. The sulfur emitted from decomposing organic matter would not affect seawater sulfate. The potential importance of this idea can be realized when we compare the amount of sulfur in organic matter with that in the sea. The oceans contain 1.3×10^{21} g S or about 1,000 billion billion tons of sulfur, nearly all as sulfate. So the amount of hydrogen sulfide you could possibly get from seawater is virtually infinite, especially since the sulfate content is being continuously replenished by rivers. By contrast, although an essential minor component of all living matter, sulfur in organic matter does not approach 1 millionth of the potential oceanic supply. This is a key idea. In the following chapters, we see why this is so important to pyrite formation and the world we live in.

Both ideas about hydrogen sulfide formation—rotting organic matter and seawater—constituted a giant leap in the understanding of the natural environment that was not appreciated at the time. It was taken for granted or, more likely, regarded as a curiosity. Sulfur was still thought to mainly originate in volcanoes. But even if the magnitude of the process was not appreciated, the idea that pyrite could be formed from something as mundane as rotting vegetable matter is a far cry from earlier theories of pyrite growing in the Earth and evolving toward gold—or even the simpler idea that metalliferous minerals only formed in volcanoes. It also contradicted any notion that minerals were formed during some single event such as biblical Creation or during the Flood. Pyrite was forming today. Crystals could be observed growing in laboratory jars containing mouse droppings.

Bacteria Hold the Key

Bacteria were first observed in 1676 by Anton van Leeuwenhoek when he accidently looked into a drop of water with his newly invented microscope. However, van Leeuwenhoek had no idea what he was seeing. All he knew was that there was a microscopic world inside a drop of water that no one had previously seen or suspected. *Bacteria* were named by the geologist

Christian Gottfried Ehrenberg in 1838.[17] The first bacteria identified were sulfur bacteria, organisms that metabolize sulfur and its compounds. John Gay, my old microbiology tutor, told me, "Dave, there are two types of bacteriologists: sulfur bacteriologists and the rest." The reasons for this were twofold. First, 99% of bacteriologists are medical microbiologists and there are few human pathogens among the sulfur bacteria. Second, sulfur bacteria are often weird and wonderful. They include the largest known bacteria. These figured strongly in early bacteriology since they can be seen with the naked eye.

The bacteria described by the early bacteriologists were sulfur-oxidizing organisms, so they did not explain the observed relationship between sulfate and organic matter. As we have seen, instead of adding oxygen to sulfur, which is an oxidation process, the transformation of sulfate to sulfide requires the removal of the oxygen. This process of oxygen removal is referred to as *reduction*.

Bacteria metabolize organic matter and often release hydrogen sulfide from the organic sulfur. In the last decades of the 19th century, there had been several reports of bacteria that produced H_2S in this manner.[18] But, as we saw earlier, this process does not explain the removal of seawater sulfate in blue muds.

Could the organic matter in the sediments have reacted with the seawater sulfate, removing the oxygen from the sulfate and producing hydrogen sulfide? And then the sulfide would react with iron salts to produce pyrite in the blue muds. This might explain the mystery of the disappearing sulfate and the appearance of pyrite observed by Murray and his colleagues on the *Challenger* expedition.

The problem with this idea is that the simple chemical reaction between sulfate and organic matter as proposed by Pepys and his followers in the 19th century does not work at Earth surface temperatures. Sulfate is an extremely stable molecule mainly because it is symmetric, with four oxygen atoms surrounding a central sulfur atom. So there is no weakness in the molecule that allows a point of chemical attack. In order to remove the oxygen inorganically we need to either heat it to temperatures above 200°C or expose it to extreme concentrations of especially potent chemicals that do not (and cannot) occur naturally at ambient temperatures in sediments. By contrast, at temperatures above 200°C the chemical process begins to become important and may contribute much of the sulfur to the sour gas associated with hydrocarbon reservoirs.

Murray and Irvine concluded that seawater sulfate could be reduced to sulfides at ambient temperatures in the presence of catalysts. They suggested that the catalysts were the bacteria involved in the decay of organic matter. The sulfide the bacteria produced reacted with ferric oxides from the surface layer of the sediments to form pyrite. Interestingly, they also reported that the seawater immediately above the muds was saturated with calcium carbonate and that carbon dioxide was a product of the bacterial decay of organic matter. It appears that by 1895 Murray and Irvine had identified the major components of the formation of sedimentary pyrite. Murray and Irvine's conclusions were substantiated in 1903 by Albert Hindrik van Delden,[19] who reached similar conclusions with respect to black, sulfidic muds from canal waters.

Van Delden had an advantage over Murray and Irving because by 1895 the Dutch microbiologist Martinus Willem Beijerinck had discovered sulfate-reducing bacteria in sewer mud from Delft (Figure 6.1).[20] Sulfate-reducing bacteria are organisms that remove the oxygen from sulfate, thereby producing sulfide. The process is analogous to removing the oxygen from the air by breathing; these bacteria take the oxygen from sulfate. The organisms are able to utilize the reduction of sulfate to sulfide as an energy source. This is chemically quite remarkable since these bacteria manage to reduce sulfate to sulfide at room temperatures and, indeed, make a living out of it: a reaction we can only manage at high temperatures or with exotic, energy-demanding chemicals. Basically they bring a complex enzyme system to bear on the reaction. Enzymes are nature's catalysts, and they enable chemical reactions to run under conditions in which they would not ordinarily work at any useful rate. The details of exactly how these organisms are able to do this were not realized until the last decade of the 20th century and, even today, major steps in the process are not entirely understood.

Beijerinck's discovery was even more significant than just explaining how sulfate was reduced in the oceans. These sulfate-reducing bacteria can exist only in environments where there is no oxygen. Indeed, oxygen is poisonous to them. We are familiar with the everyday surface environment of the Earth—it has plenty of oxygen, which we freely breathe. We are happily unaware that the most of the Earth's environment is in fact oxygen-free. Oxygen in the air is used up greedily by organisms that need oxygen to live. Any oxygen that is left over reacts with decomposing organic matter and is removed. So if you go just a few centimeters below the surface, especially in any lake or coastal strand, you will quickly find an environment that is

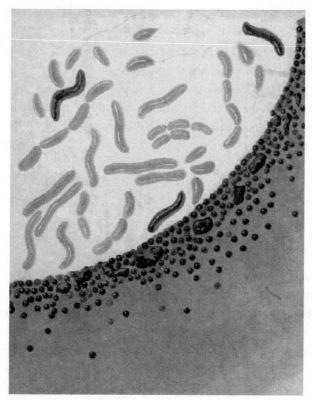

FIGURE 6.1. Martinus Beijerinck discovered sulfate-reducing bacteria in 1895, and this beautiful, microscopic sketch of the original organisms was drawn by his sister Henriëtte. Courtesy of the Beijerinck Archives, Delft.

oxygen-free. What Beijerinck showed was that, rather than being sterile and free of life, these anoxic environments are potentially teeming with a universe of organisms that can exist only in oxygen-free systems.

Beijerinck called the first sulfate-reducing bacterium he isolated *Spirillum desulfuricans*. Under the light microscope they appear as tiny squirming commas, each no more than about one-millionth of a meter in size. It is popular to compare the size of small things to the thickness of a human hair. Well, you would get about 100 across a hair if they were lined up head to tail. They can really only be studied in detail with the help of an electron microscope.

Spirillum desulfuricans turned out to include a large number of different organisms, and these were reclassified in the late 20th century. The most common was renamed *Desulfovibrio desulfuricans* when I first worked with them. This has since changed to *Desulfovibrio vulgaris*. These are more or less the same organism. *Vibrio* refers to the comma-like shape and *vulgaris* to the fact that it is common—and it is extremely common. The original strain of *Desulfovibrio desulfuricans* was called *Hildenborough* since this is the name of the ditch outside the laboratory where Mary Adams found it. Mary Adams was the assistant to the great John Postgate, one of the doyens of sulfate-reducing bacteriology in the mid-20th century. Mary had magic fingers as far as growing bacteria were concerned, in the same way as some people have green thumbs for plant cultivation. I have followed Mary around the laboratory doing exactly the same as her: a teaspoonful of this and a tablespoonful of that to make up a growth medium for the bacteria. All her bacteria grew and all mine died.

Desulfovibrio vulgaris strain Hildenborough has become a standard sulfate-reducing bacterium because it is so tolerant. It can even withstand some oxygen in the atmosphere for a short time, so it is an ideal sulfate-reducer to give to a student. I took a picture of it with my electron microscope, which is currently on display in the Science Museum, London. It is a small cell with a long tail, called a flagellum (Figure 6.2). The flagellum allows it to swim so it can dive down into the safety of the anoxic sediment if the oxygen levels get too high. I should make a confession here. These specimens were prepared by putting a drop from a *Desulfovibrio* culture onto a collodium film covering an electron microscope grid. The water was evaporated and the organisms were coated in tungsten. The reason for this was to make them opaque to the electron beam. However, the result of this mistreatment was a mass of cells and detached flagella. If you look closely at Figure 6.2 you can see that the flagellum is overlapping the cell wall. The image was actually a happenstance: I found one cell that had a flagellum nearly in the right place for the image.

The second genus of sulfate-reducing bacteria isolated was *Desulfotomaculum*. This is an entirely different organism. Its name was coined by John Postgate's father, a classics don at Cambridge. John showed his father a photograph of the organism and asked his father to suggest a name. His father thought that it looked like a sausage, so it became *Desulfotomaculum*, "sulfate-reducing sausage" in Latin. I have often used this as an example to assuage the concerns some students have about the long Latin names of some organisms. The English translation is usually banal and sometimes amusing. *Desulfotomaculum* can get relatively large, more than ten times as long as *Desulfovibrio*. It usually has many flagella all over the surface of the cell and tends to wallow in thick mud.

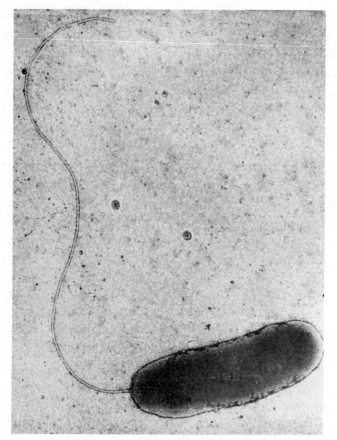

FIGURE 6.2. Electron micrograph of *Desulfovibrio desulfuricans*. The single-celled organism is about one-millionth of a meter in length and has a flagellum (tail) for movement.

The original species would only grow at relatively high temperatures too. It needed about 45°C and could grow in quite hot 60°C water. The other aspect of this organism is that it produces endospores, tiny, tough capsules of DNA that resist heat, aridity, and even chemical attack. These endospores make sterilizing a *Desulfotomaculum* culture difficult since the usual autoclave methods do not work. *Desulfotomaculum* belongs to an entirely separate phylum of bacteria from *Desulfovibrio*. Indeed it may be a relict of a phylum from which all modern bacteria have evolved.

I describe *Desulfotomaculum* in some detail in order to suggest the extraordinary variety of the sulfate-reducing bacteria. There are now more than 140 species belonging to 35 distinct genera in 4 different phyla. One of these phyla is not actually a bacterial phylum. It belongs to the Archaea, a domain of single-celled life discovered in the late 20th century that is genetically closer to us than to the bacteria. These Archaea are interesting since they mainly grow at very high temperatures, above 80°C, and are found in the boiling waters of geothermal springs and deep-ocean vents. These temperatures would kill most other life forms.

The result of this diversity of sulfate-reducers is that they are found worldwide in all environments where there is a little sulfate and no oxygen. The high-temperature varieties are matched by low-temperature species that only grow at subzero temperatures in saline lakes. The species enjoying Earth-surface conditions are complemented by others that require huge pressures of the deep oceans or the Earth's crust to grow. The sulfate-reducers are major components of the deep biosphere of the Earth, which was discovered by my colleague John Parkes and his team in the 1990s.[21] This deep biosphere extends down over 1,000 m into the crust and has resulted in our measure of the Earth's biosphere being 10% greater than was previously thought.

Bacteria and Pyrite

We now have defined the process for producing pyrite in sediments. Sulfate-reducing bacteria reduce the sulfate in natural waters, such as the ocean, to hydrogen sulfide. The sulfide then reacts with iron salts in the environment to form pyrite. This sounds possible, but it still needed to be proven. I ran an extensive and systematic program in the early 1960s that examined this theory.

The idea that bacteria can make minerals was revolutionary when I started my research. I wrote a dissertation on bacterial pyrite that not only gained me a research scholarship but also a £10,000 grant from the now-defunct Department of Scientific and Industrial Research, the UK equivalent of the National Science Foundation. With this enormous sum, equivalent to US$300,000 today, I built a geomicrobiology laboratory in Imperial College. The aim of the project was to synthesize sulfides like pyrite using bacterially produced sulfide. At that time it was not known if the bacterial sulfide would react differently from inorganic sulfide.

This first geomicrobiology laboratory caused great interest among the mining engineers and geologists in the Royal School of Mines at Imperial College, who were fascinated by these tiny organisms that could produce minerals like pyrite. They would come to the laboratory and look in the incubator. "What do you feed them on?" they would ask.

I fed them a mixture of lactate and yeast extract, but the miners were not familiar with either of these substances. Lactate is of course a constituent of cheese, and yeast extract is the main constituent of a black, savory spread called Marmite, a by-product of beer brewing in the United Kingdom. So I said that I fed them on cheese and Marmite, which seemed to satisfy them, since they were generally keen on cheese and Marmite themselves.

"So how do you know when they're hungry?" they persisted.

I recount this to illustrate how revolutionary the idea that bacteria could make minerals like pyrite was in the 1960s. It was featured in national newspapers in the United Kingdom and broadcast on the BBC both in the United Kingdom and worldwide.

In fact, like many ideas, it was not new. The Russian bacteriologist B.L. Issatchenko had originally reported iron sulfides within the cells of sulfate-reducing bacteria in 1912. Issatchenko had been a pupil of Beijerinck and was later to serve as the director of the great Winogradsky Institute in Moscow for many years. He published his study in the *Annals of the Imperial Botanical Garden of St. Petersburg*. The copy I saw in the London University Library was beautifully printed and includes hand-colored ink sketches of sulfate bacteria with tiny pyrite crystals glistening in their cells.

Eugene Thomas Allen was one of the pioneers in geochemistry in the United States and an early member of the great Carnegie Institute of Washington. His views on the relationship between microorganisms and sedimentary pyrite formation were therefore very influential. In 1912 he and his colleagues discussed Beijerinck's bacteria in some detail.[22] They cited a coworker, C.A. Davis, as reporting that he always found hydrogen sulfide in marine peat bogs into which seawater had ingressed. The important thing here is that seawater is enriched in sulfate, and sulfate-reducing bacteria need sulfate to function. Even so, Allen concluded that microorganisms were not generally responsible for the large quantities of pyrite in sediments since "they were only found near the surface."

Fortunately, US sulfide research took a new direction with the appointment of Lourens Baas Becking, a graduate of the Delft school, at Stanford in 1923 and his hiring of Kees van Niel, a student of van Kluyver at Delft, to Stanford University's Hopkins Marine Station in 1929. Many pioneering US microbial ecologists were trained at the Hopkins Marine Station and went on to establish their own research programs at institutions throughout the United States. Baas Becking was one of the great pioneers of geomicrobiology and is famous for defining the chemical limits of microbiological activity. He had a research institute named after him at the Australian National University and Geological Survey in Canberra in the 1970s, which did much work to progress early geomicrobiology. I worked as a research fellow there in 1974.

By this time, the formation of iron sulfides had become a diagnostic feature for identifying sulfate-reducing bacteria in cultures. The organisms produce sulfide, which reacts with iron in the culture medium to produce a characteristic black precipitate of iron sulfide. I made detailed studies of this precipitate. It seems that the cells in healthy bacterial cultures are mainly free of iron sulfide.[23] However, as the environmental conditions deteriorate and the organisms become less vigorous, the individual cells become coated with iron sulfide (Figure 6.3).

It is not known whether this confers some advantage on the organisms, but it seems doubtful. Microorganisms have developed extensive defensive measures, usually in the form of the production of large amounts of extracellular proteins, to protect themselves from iron sulfide precipitates. The process can result in huge amounts of gunk filling a test tube of bacteria and producing far more organic matter than is contained in the organisms themselves. We discovered that a possible reason for this is

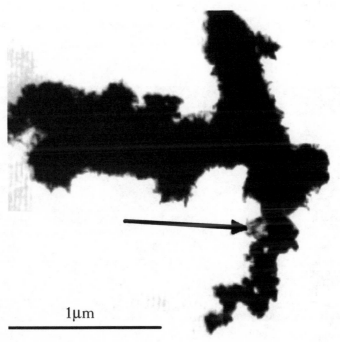

1μm

FIGURE 6.3. Electron micrograph of a sulfate-reducing bacteria coated with iron sulfide. A flagellum is arrowed. The scale bar is 1 micrometer or a millionth of a meter. Photo Rolf Hallberg.

that nanoparticulate iron sulfides attack DNA, causing breaks in the molecule and ultimately resulting in unraveling the double helix. The DNA ceases to function in this uncoiled state. This is not a good thing, and the bacteria expend large amounts of energy in counteracting it. It seems probable that the iron sulfide coatings of sulfate-reducing bacteria are not characteristic of vigorous, healthy populations.

L.P. Miller[24] synthesized a number of metal sulfides using sulfate-reducing bacteria in 1950. Lourens Baas Becking[25] continued the study. My study showed that there were no detectable differences between these bacterially produced metal sulfides and those produced in inorganic experiments. I have to add a caveat here: I could find no differences within the precision of the analytical technology available at the time. This has advanced immensely since the early 1960s, and it would be worth running the program again. Since that time a number of organisms have been discovered that deliberately produce iron sulfides within their cells. These include the magnetotactic bacteria, a group of organisms that synthesize chains of magnetic iron sulfide within their cells, apparently enabling them to line up with the Earth's magnetic field.

A snail that lives on the bases of black smokers at a depth of 2,400 m in the Indian Ocean has a foot armored by scales of iron sulfides including pyrite (Figure 6.4).The snail's pyrite armor is actually produced by a symbiotic community of bacteria, including sulfate-reducers.

FIGURE 6.4. A snail with its foot armored with overlapping pyrite scales living near a deep-sea hydrothermal vent at a depth of 2,400 m in the Indian Ocean. Photo by Anders Warén.

Pyrite Isotopes

The atomic weight of sulfur is 32.065. Since individual atoms are made up of discrete particles, it is not possible for a single atom of sulfur to have a weight that is not a whole number. The reason the atomic weight of sulfur is not a whole number is that it is the average weight of a sample of different sulfur atoms with weights between 32 and 36. These are isotopes of sulfur, but, unlike radioactive isotopes, they do not spontaneously decompose. They are quite stable, and they can be separated or fractionated by chemical reactions. Isotopes are simply atoms that have identical outsides but different insides, as Sir Frederick Soddy explained in his 1922 Nobel Lecture.

The average atomic weights of sulfur samples vary according to their origin. Generally, the lighter sulfur isotopes react faster chemically than the heavier isotopes. So in a chemical reaction that involves a number of steps, such as the bacteriological reduction of sulfate to sulfide, the lighter sulfur isotope is concentrated in the product sulfide compared with the reactant sulfate. The nice thing about this is that this fractionation provides a chemical signature for bacterial sulfate reduction.

Pyrite is sulfur-rich, and the precise analysis of the isotopic mass of pyrite-sulfur is fairly easy. Technically all that is required is to take a sample of pyrite, burn it, and pass the resultant sulfur gas through a mass spectrometer. It is not even necessary to measure the absolute concentration of each sulfur isotope. The relative concentrations are sufficient, and these can be measured to a far greater degree of precision. Sulfur isotopes in pyrite were therefore among the first stable isotopes to be measured in 1949, just two years after Harold Urey published his original Nobel Prize-winning paper on the theory of stable isotopes. The composition of sulfur isotopes in pyrite is used to determine if the sulfur in the pyrite has been produced by bacterial processes.

Measurements of the sulfur isotopic compositions of sedimentary pyrite therefore have provided proof of the role of bacterial sulfate reduction in the production of pyrite in sediments. Pyrite formed from bacterial sulfate reduction is enriched in the lightest sulfur isotope. Interestingly, pyrite formed from volcanic sulfur is not enriched in the lightest isotope since the production of the reactant sulfur is at high temperatures, which do not induce significant isotope fractionations. The consequence of being able to prove the bacterial origin of pyrite-sulfur is that we are also able to prove the volcanic origin. As discussed in Chapter 8, this gives a method of estimating the relative importance of bacterial and volcanic processes in global pyrite formation.

Recent technological advances have been made in both the machines used to measure stable isotopes and the chemistry of preparation of samples for analysis. This has enabled not only increased precision of the more abundant isotopes of sulfur but also better measurements of the least abundant stable isotopes. The mathematics of the fractionation processes are very robust and have been used to acquire information on present and past biogeochemical cycles and predict the future of the Earth's surface environment. The enhanced precision of measurements has resulted in surprising information about the early history of the Earth's atmosphere, the detail of the bacterial reduction process, and even greater discrimination between bacterial and abiological sulfate reduction processes. These are discussed in more detail in Chapter 9.

The technological advances in stable isotope measurements have also enabled the isotopes of iron in pyrite to be analyzed. The principles of isotope fractionation are the same as for sulfur, but the subject is still in its infancy. Measurements of the iron isotopic compositions of ancient sedimentary pyrite have revealed significant variations in the past. However, the reasons for these fractionations are not well understood yet. Our experimental work has shown that the fractionations may be controlled at low temperatures by the reaction rates, and this would mean that there is no unique solution to interpreting these variations. Even so, the potential of combining sulfur and iron isotope measurements of pyrite is very exciting.

Notes

1. Sulfur appears to be related to ancient Indo-Aryan roots through the Sanskrit *sulvere, sulvari,* or *shulbari* (or *zailavara* in a phonetic transliteration). *Shulbari* may derive from *ari* (enemy) and *shulba* (copper) and perhaps refers to the reaction between sulfur and copper to blacken it, or this may be chemical poetic license. The Latin word was inherited from the Etruscan.

2. For example, C.W. Mitchell and S.J. Davenport. 1924. Hydrogen sulphide literature. *Public Health Reports,* 39:1–13.

3. T.O. Bergman. 1784. *Physical and Chemical Essays* (Edinburgh: Mudie), 433pp.

4. J.J. Berzelius. 1819. *Essai sur la théorie des proportions chimiques et sur l' influence chimique de l' électricité* (Paris: Méquignon-Marvis).

5. T.L. Guidotti. 1996. Hydrogen sulphide. *Occupational Medicine-Oxford*, 46:367–371.

6. F. Hoppe-Seyler. 1863. Einwirkung des Schwefelwasserstoffgases auf das Blut. *Zentralblatt für die medizinischen Wissenschaften*, 1:433–434.

7. See J. Strickland, A. Cummings, J.A. Spinnato, J.J. Liccioe, and G.L. Foureman. 2003. *Toxicological Review of Hydrogen Sulfide*. 635/R-03/005 (Washington, DC: US Environmental Protection Agency), 67pp.

8. W. Skey. 1893. On the nature of stinkstone (anthraconite). *Transactions and Proceedings of the Royal Society of New Zealand*, 25:379–380.

9. For example, D. Rickard, M.Y. Willdén, N.E. Marinder, and T.H. Donnelly. 1979. Studies on the genesis of the Laisvall sandstone lead-zinc deposit, Sweden. *Economic Geology*, 74:1255–1285.

10. D. Rickard and J. Morse. 2005. Acid volatile sulfide. *Marine Chemistry*, 97:141–197. The amount of pyrite in these sediments is insufficient in itself to color the sediments blue-gray. The coloration seems to stem from the absence of red and yellow oxidized iron minerals, black iron sulfides, and organic matter, although, as noted further on, Murray and Renard thought it was caused by a combination of organic matter and pyrite particles.

11. J. Murray and A.F. Renard. 1891. *Deep-Sea Deposits: Report on the Scientific Results of the Exploring Voyage of H.M.S. Challenger (1873–1876)* (Edinburgh: John Menzies & Co.).

12. J.Y. Buchanan. 1890. On the occurrence of sulphur in marine muds and nodules, and its bearing on their mode of formation. *Proceedings of the Royal Society of Edinburgh*, 18:17–39.

13. J. Murray and R. Irving. 1895. On the chemical changes which take place in the composition of the sea water associated with blue muds on the floor of the ocean. *Transactions of the Royal Society of Edinburgh*, 37:481–507.

14. W.H. Pepys. 1811. XVIII: Notice respecting the Decomposition of Sulphate of Iron by Animal Matter. *Transactions of the Geological Society of London*, Series 1, 1:399–400.

15. R. Bakewell. 1838. *An Introduction to Geology: Intended to Convey a Practical Knowledge of the Science and Comprising the Most Important Recent Discoveries; with Explanations of the Facts and Phenomena which Serve to Confirm or Invalidate Various Geological Theories* (London: Longman, Orme, Brown, Green & Longmans), 657pp.

16. See, for example, G. Bischoff. 1832. Die bedeutung der mineralquellum und der gasekhalation bei der bildung und veränderung der endoberfläche dargestellt nach geoghastischen beobachtungen und chemische untersuchungen. *Schweiggers Journal of Chemistry and Physics*, 64:377–409. A.P. Brown. 1894. A comparative study of the chemical behavior of pyrite and marcasite. *Proceedings of the American Philosophical Society*, 33:225–243. A. Gautier, 1893. Formation des phosphates naturels d'alumine et de fer. Phénomènes de la

fossilisation. *Comptes Rendus hebdomadaires des séances de l'Academie des sciences,* 116:1491–1496.

17. C.G. Ehrenberg. 1838. *Die Infusionsthierchen als volkommene Organismen. Ein Blick in das tiefere organische Leben der Natur* (Leipzig: Voss).

18. For example, P. Miquel. 1889. Biogénese de l'hydrogène sulfuré. *Annales de Micrographie,* 1: 323–364.

19. A.H. van Delden. 1903. Beitrag zur kenntnis der sulfatreduktion durch bakterien. *Zentralblatt für Bakteriologie Parasitenkunde Infektionskrankheiten und Hygiene II Abteilung-Naturwissenschaftliche-Mikrobiologie der Landwirtschaft der Technologie,* 11:81–94.

20. M.W. Beijerink. 1895. Über Spirillum desulfuricans als Ursache von Sulfatreduktion. *Centrablatt für Bakteriologie,* Abt. I:1–9, 49–59, 104–114.

21. R.J. Parkes, B.A. Cragg, S.J. Bale, et al. 1994. Deep bacterial biosphere in Pacific-ocean sediments. *Nature,* 371:410–413.

22. E.T. Allen, J.L. Crenshaw, J. Johnson, and E.S. Larsen. 1912. The mineral sulphides of iron with crystallographic study. *American Journal of Science,* 33:169–236.

23. Iron sulfide nanoparticles have been observed scattered in the protoplasm of some sulfate-reducing bacteria; see D. Rickard. 2012. *Sulfidic Sediments and Sedimentary Rocks* (Amsterdam: Elsevier), 801pp.

24. L.P. Miller. 1950. Formation of metal sulfides through the activities of sulfate-reducing bacteria. *Contributions from the Boyce Thompson Institute for Plant Research,* 16: 85–89.

25. L.G.M. Baas Becking and D. Moore. 1961. Biogenic sulfides. *Economic Geology,* 56:259–272.

7

Acid Earth

THE ATMOSPHERE AND much of the rivers, lakes, and oceans of the Earth are oxygenated. Any pyrite that comes into contact with these environments becomes unstable and breaks down. The process is called *oxidation*. It is an exothermic process and, as described in Chapter 5, this process was thought to heat the Earth. It is the opposite of *reduction*, which we discussed with regard to the microbial formation of sulfide from sulfate in Chapter 6.

The counterintuitive concept important here is that oxidation is a chemical process that does not necessarily need oxygen. This idea—that you can oxidize things in the absence of oxygen—is one that most natural scientists are aware of but that they need a couple of nudges occasionally to remind themselves about.[1] This means that pyrite oxidizes not only in oxygenated environments—although that is what we are most familiar with—but also in oxygen-free environments.

Among the products of pyrite oxidation are large quantities of acid. Although this happens naturally during rock weathering, the intervention of humankind has led to an enormous increase in the exposure of pyrite to the atmosphere. This has produced contamination of the atmosphere, groundwater, and watercourses on a regional scale. It has also increased the amount of uncontrolled coal burning in coal seams, coal mines, and coal waste tips worldwide, making whole towns uninhabitable and laying waste to large areas.

In this chapter I consider in more detail what exactly the process of pyrite oxidation is and how it affects the Earth's environment today, as well as the problems it stores up for humanity in the future.

Chemical Pyrite Oxidation

In chemical terms, oxidation does not mean just the addition of oxygen. Oxidation is a reaction that involves the removal of one or more electrons from a compound because of a chemical reaction. One of the most familiar oxidation reactions is combustion, where substances burn in air to produce heat. The way to put out such a fire is to restrict oxygen access using a chemical foam or fire blanket. Since this reaction with oxygen was the best known, the process was called originally called *oxidation*. However, oxygen is just one of the substances that can capture electrons from iron and sulfur. There are many other oxidizing compounds.

Just as elements exist as isotopes with differing atomic weights, elements also exist with different numbers of electrons (Figure 7.1). Continuing Professor Soddy's analogy from the previous chapter, these are atoms with the same insides but different outsides. The various numbers of electrons denote the oxidation state of the element or compound, since the fewer electrons an element has, the more it can oxidize—that is, grab electrons from—other compounds. The outside of an atom of metallic iron has twenty-six electrons. However, iron commonly exists in water with either twenty-three or twenty-four electrons. Iron with twenty-four electrons is called *ferrous* iron and often forms compounds with a green

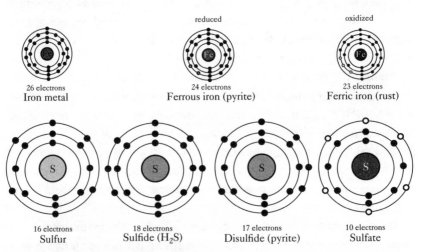

| reduced | oxidized |

26 electrons
Iron metal

24 electrons
Ferrous iron (pyrite)

23 electrons
Ferric iron (rust)

16 electrons
Sulfur

18 electrons
Sulfide (H$_2$S)

17 electrons
Disulfide (pyrite)

10 electrons
Sulfate

FIGURE 7.1. Simplified solar system–type diagrams of oxidized and reduced iron and sulfur atoms showing the different numbers of electrons. The more oxidized forms have fewer electrons; oxidation is simply the process of losing electrons and is not necessarily related to oxygen.

color; iron with twenty-three electrons is called *ferric* iron and commonly forms red-colored compounds. Ferric iron has fewer electrons and is oxidized relative to ferrous iron, which has more electrons. Ferric iron is the form of iron in rust, which is the familiar product of iron oxidation. Pyrite contains ferrous iron and, when pyrite oxidizes, the ferrous iron loses an electron and ferric iron, with twenty-three electrons, is produced. This ferric iron precipitates in water as rust.

Sulfur is more complex and can exist with multiple oxidation states usually containing between ten and eighteen electrons. Solid sulfur has sixteen electrons. Sulfur, with eighteen and seventeen electrons, is the least oxidized—or most reduced—and these forms of sulfur are called *sulfides*. Pyrite is an iron sulfide and contains sulfur with seventeen electrons. When it reacts with oxygen it loses these electrons, forming an oxygen compound with just ten electrons called *sulfate*, SO_4. When pyrite oxidizes, the end product of the oxidation of the sulfur is sulfate.

Of course, you do not get an electric shock when touching an oxidizing pyrite crystal. The oxidation reaction produces electrons, and, unless these are somehow used up, the result would be an electrical imbalance. Therefore every oxidation reaction that produces electrons must be counterbalanced by a reduction reaction that consumes electrons. The two halves of the process are often called *redox reactions*, a neat, portmanteau term that encapsulates the idea that they involve both oxidation and reduction processes. Redox reactions involve the exchange of electrons between substances. In the case of the oxygen reaction with pyrite, the counterbalancing reducing reaction is the capture of the electrons from the sulfur and iron by oxygen. The product is water, which is an oxide of hydrogen: H_2O.

This balance of reduction and oxidation reactions extends to oxygen-breathing organisms like us. Our reduction reaction is the reaction of the oxygen to water, and this is counterbalanced by the oxidation reaction, which is the formation of carbon dioxide from organic matter. So as you sit quietly reading this book you are carrying out a redox reaction. You are breathing in oxygen from the air and oxidizing the steak you had for dinner to carbon dioxide, which you gently exhale. The net energy balance of this process gives you the energy to turn the page over and gives your brain the power to understand oxidation and reduction reactions.

In the case of the oxidation of pyrite by ferric iron, the pyrite sulfur is oxidized to sulfate and water is reduced to hydrogen ions, which are the basic constituents of acids. In this case, the ferrous iron in the pyrite is merely released into the solution as dissolved ferrous iron with no oxidation or reduction.

The breakdown of pyrite in the Earth's surface environment is quite complicated. It involves the oxidation of both iron and sulfur. Although iron exists in only two oxidation states, sulfur can occur in eight. It gets worse—because oxidized ferric iron is itself an oxidizing agent of sulfide. So there is a positive feedback. As pyrite oxidizes with oxygen, more ferric iron is formed, and this in turn oxidizes the pyrite and so on.

These two processes describe the basic reactions involved in the oxidation of pyrite in the natural environment. One reaction is the oxidation of iron and sulfur with oxygen, usually derived from the air or from oxygen dissolved in water. The second reaction, with ferric iron, does not require oxygen. It is therefore described as the *anoxic pyrite oxidation pathway*. Of course, in most cases oxygen may be needed to produce ferric iron from ferrous iron in pyrite in the first place. However, we can envisage situations where the ferric iron is produced in an oxygenated environment and transported to an anoxic system where it will continue to oxidize pyrite.

The acid produced in pyrite oxidation is a strong mineral acid—sulfuric acid—which results from the dissolution of sulfate in water. As discussed in Chapter 2, sulfuric acid is the most abundant manufactured chemical, and its production is so dominant that it can be used as a measure of economic activity. The production of acid from the oxidation of pyrite is quite dramatic. For every two molecules of pyrite oxidized, twice as many molecules of sulfuric acid are produced by the oxygen reaction. The oxidation of pyrite by ferric iron is even more efficient as an acid producer. The equivalent of sixteen molecules of acid are produced for each molecule of pyrite oxidized. The result is interesting since, by contrast with many other minerals (including many sulfides), pyrite does not dissolve easily in mineral acids. An oxidizing acid is necessary to get pyrite to dissolve. This means that the oxidation of pyrite produces mineral acid, which can then go on to dissolve many other minerals, including other metal sulfides, releasing other, more toxic elements like mercury and arsenic into the environment.

The processes involved in the oxidation of pyrite have been primary research targets for the past half-century. The problem has been in unraveling these interdependent reactions. We are familiar with pyrite lasting a considerable amount of time in air: the beautiful pyrite crystals would not exist unless the reaction of pyrite and air was slow. In fact, the pyrite surface evolves a molecular layer of iron oxide in air that protects it from further oxidation. However, in water the oxidation products at the pyrite surface become soluble and may be transported away. The result is the continued oxidation of pyrite. Recently my colleague George Luther of the University of Delaware has given good chemical reasons why pyrite in sterile low-temperature aqueous environments can sometimes appear to oxidize slowly. We can see that the oxidation of iron involves the removal of just one electron. By contrast, the oxidation of pyrite-sulfur involves the removal of seven electrons. If the removal of each electron involves a separate chemical reaction, the process may be quite slow unless it is biologically catalyzed.

Microbial Pyrite Oxidation

It is obvious that the theoretical chemistry of pyrite oxidation does not agree with what is actually observed in nature. Pyrite placed in a bucket of water oxidizes rapidly rather than slowly, and the cause of this rapid oxidation is catalysis by bacteria. The oxidation of sulfur and iron are exothermic, energy-producing processes, and it is not surprising that life has found a way of latching on to this free meal.

One of my earliest jobs was trying to improve the rate of pyrite oxidation by bacteria. I cultured sulfur- and iron-oxidizing bacteria in what was called an air-lift system. This was a glassware apparatus that used bubbled air to mix the pyrite slurry and to provide oxygen for the bacteria. It had the additional advantage that bubbling air through the pyrite slurry churned the pyrite particles and exposed new pyrite surfaces for further reaction. My specific task was to see if the bacteria could be trained to tolerate higher amounts of copper. Copper can be an effective bactericide and was often produced by acid attack in both industrial and natural environments where pyrite was oxidized. We noted in Chapter 3 that massive pyrite ores often contain economically important copper concentrations, so it was important to find out if the bacteria could be trained to tolerate high dissolved copper concentrations. In fact, they could. A bacterial culture that started by almost totally dying off even in trace copper concentrations always had a few cells that managed to thrive and prosper. Reculturing the survivors produced bacterial populations with increasingly high copper tolerances. This is a cheap form of genetic engineering: nowadays it might be approached by splicing copper-resistant genes into the bacterial DNA.

As mentioned in Chapter 6, Christian Gottfried Ehrenberg coined the word *bacteria* in 1838 as a name for a subset of microorganisms with rod-shaped forms. The word *bacterium* is derived from the Greek *bakterion* or "small rod" and exactly describes one of the major groups of bacteria involved in pyrite oxidation. These small rod-shaped organisms are about one-thousandth of a millimeter in length (Figure 7.2). Different species oxidize sulfur to sulfate and ferrous iron to ferric ion. They attach themselves to the pyrite surface and extract the energy from iron and sulfur oxidation for their metabolism.

One of the characteristics of these bacteria is their extreme acid tolerance. Indeed, some varieties will grow only in very acidic solutions. This is a neat evolutionary trick. The bacteria catalyze the production of acid through the oxidation of pyrite and create an environment that it is extremely hostile to most other organisms. The solutions they produce and tolerate can be extremely acidic—enough to produce chemical burns on human skin.

The bacteria catalyze the rate of pyrite oxidation by extraordinary amounts. Pyrite oxidation with oxygen can be several hundred times faster in the presence of bacteria. The rate of oxidation with ferric iron

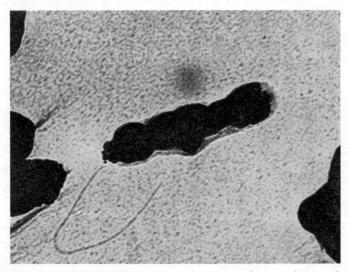

FIGURE 7.2. A rod-shaped sulfur-oxidizing bacterium from acid mine drainage. The black coloration is iron oxide from pyrite oxidation. The bacterium is about one-millionth of a meter long and has a flagellum for motility. Photo Rolf Hallberg. From D. Rickard and R.O. Hallberg. 1973. De Små gruvarbetarna i Falun. *Svensk Naturvetenskap*, 102–107.

becomes dominant in more acid solutions and can be 1,000 times faster in the presence of sulfur- and iron-oxidizing bacteria. Thus pyrite oxidation and concomitant acid production in the environment can be closely approximated as an entirely bacterial process.

Many organisms have evolved different strategies to benefit from the energy available from pyrite-sulfur oxidation. We have already seen that the process can be carried out by ferric iron in the absence of oxygen, but bacteria can couple other oxidizing agents to the pyrite oxidation. The one that disconcerted the scientific community the most was the nitrate route. Environmental scientists were familiar with pyrite oxidation on land in contact with our present oxygen-rich atmosphere, where the bacteria commonly need oxygen to prosper and the media involved are highly acidic, ferric iron solutions. In the nitrate system, bacteria couple the reduction of nitrate, NO_3, to sulfur oxidation. In the same way the sulfate-reducing bacteria remove oxygen from sulfate, SO_4, these bacteria remove oxygen from nitrate, NO_3, but then they use this to oxidize sulfur. The important thing about this is that nitrate reduction does not require acidic conditions. The result is that the oxidation of pyrite occurs not only in the absence of oxygen but also in water that is not acidic, such as normal seawater.

Acid Rivers

The Rio Tinto mining district in Spain gets its name from the red river that courses through one of the greatest concentrations of pyrite on the planet, perhaps comprising several billions of tons of the mineral. The red color is caused by iron ochers released by the oxidation of pyrite (Figure 7.3).

The mines along the Rio Tinto have been worked for thousands of years: Iberian and Tartessian tribes started the mines around 3000 BCE and mining was continued intermittently by Phoenicians, Greeks, Romans, Visigoths, and Moors to the modern Europeans of the present day. The combined effects of rainwater and oxygen and the gentle attentions of myriads of sulfur- and iron-oxidizing bacteria have resulted in pyrite oxidation and the production of an acidic river.

Pollution of the Rio Tinto has been recognized for a very long time. During the 16th century the mines were not being worked and King Felipe II of Spain sent the Diego brothers to review the potential mineral wealth of the province. The priest, Diego Delgado, sent his report in the form of a letter to King Felipe II in 1556[2] in which he noted that the river was called the Red River (*Rio Tinto*) because "it springs from vitriol." Delgado

FIGURE 7.3. The Corta Atalaya open pit in the Rio Tinto district, Spain. The massive pyrite body is exposed in the orange-colored zone on the far wall (rendered white on this image), which has resulted from pyrite oxidation.

reported that the river was extremely acidic—vitriol being essentially iron sulfate in sulfuric acid. Delgado wrote,

> In this river there is no type of fish nor living creature and neither people nor animals drink these waters. Another property of the water is that if iron is put into it, some it dissolves in a few days; I proved this myself by putting a living frog into the river which died without leaving the water.

Delgado had listed many of the characteristics of the acid mine drainage that is created by pyrite oxidation wherever rivers intersect pyrite deposits. The production of acid water from pyrite mines was well known in Europe from ancient times and Galen, for example, reported vitriol being collected from the mines in Cyprus in the first century CE. Delgado, however, went further and described, with a neat experiment, the effect of these acidic waters on life.

Acid mine drainage from the Great Copper Mountain at Falun in Sweden has been continuous since the mine opened in 1100 CE. It received a boost in 1687 when the mine collapsed into the open pit, allowing air and water to penetrate deeper into the pyrite rubble. As shown in Figure 5.5 in Chapter 5, the precipitates from the acid mine waters are multicolored mixtures of iron, copper, calcium, magnesium, zinc, and lead sulfates in a mass of red iron ochers. Acid mine waters do not only contain iron and acid; the acidic solutions dissolve other elements associated with the pyrite such as copper, zinc, lead, and cadmium, and non-metals such as arsenic and selenium, all of which are highly toxic in quite low concentrations.

In the same year as Delgado wrote his report to the King of Spain, the German mining geologist, Agricola, wrote in his great monograph on mining *De Re Metallica,*

> Further, when the ores are washed, the water which has been used poisons the brooks and streams, and either destroys the fish or drives them away. Therefore the inhabitants of these regions, on account of the devastation of their fields, woods, groves, brooks and rivers, find great difficulty in procuring the necessaries of life, and by reason of the destruction of the timber they are forced to greater expense in erecting buildings. Thus it is said, it is clear to all that there is greater detriment from mining than the value of the metals which the mining produces.

Agricola was describing the situation in the Bohemian mines in late medieval times and recording one of the key arguments of people opposed to mining. The same arguments echo down the centuries and are heard today from planners, environmentalists, green activists, and NIMBYs, locals who simply say "Not In My Back Yard."

A resident took me aback in the town of Jerome, Arizona, which has become an artists' colony. He said that Jerome would be ideal if it was not for the mine. I was surprised since the only reason for the original establishment of the township of Jerome in the middle of nowhere between Flagstaff and Phoenix was the great United Verde mine, which extracted copper from a massive pyrite deposit (Figure 7.4). The United Verde mine worked two giant massive pyrite bodies, probably amounting to over half a billion tons of ore, for its copper, lead, and zinc content. The deposits are the largest massive pyrite deposits in the United States and compare with the Rio Tinto ores in size—although the copper grade is far higher.

FIGURE 7.4. Exposed wall of the open pit of the giant United Verde massive pyrite deposit, Jerome, Arizona, in 1989, some thirty-five years after the mine was closed—showing multicolored rocks resulting from the oxidation of the massive pyrite. The massive size of the orebody is apparent when you consider that the small adit center left is at least man sized.

The deposit had a checkered history: the deposit is so pyrite-rich that it spontaneously caught fire and burned for more than twenty years from 1894 until it was finally extinguished in 1916.

The problem the modern residents of Jerome had was the acid mine drainage that flowed from seeps and a mine tunnel into the local creek and the Verde River. The water was azure blue, being both acidic and copper-rich. Local children would throw nails and bits of cars into it and watch them turn into copper. The mine was owned by the Phelps Dodge corporation, which spent several million US dollars on controlling surface and groundwater discharges from the mine after it was taken to court in 2003 by the US Environmental Protection Agency (EPA). Subsequently Phelps Dodge was taken over by the giant Freeport organization, which spent further millions of dollars in the remediation work. Finally, in 2013 it was reported that the azure water had stopped flowing into the creek. The total cost of the remediation work is unknown but appears to be of the order of tens of millions of US dollars.

The Falun, Jerome, and Rio Tinto situations are examples of how mining itself can cause acid mine waters. A more acute problem is with old abandoned mines. Mines usually go down beneath the water table and they are kept relatively dry during mining by continuous pumping. The water being pumped out of the mine will often be very acid but it can, in principle at least, be treated. All this ceases when the mine is abandoned. The pumps cease and the mine floods and the acidic mine waters seep into the groundwaters.

However, the more common problem in mining today is caused by the wastes from the mining operation. Obviously, if a metal ore has a concentration of 10% metal then 90% is waste rock. It gets worse. No process is 100% efficient and therefore not all of the 10% metal is actually won as metal: the process also wastes a percentage. Finally, the mine works to a certain grade of metal that is profitable at the time. This means that the waste rock may contain substantial concentrations of metal. Indeed, it is said that technological advances and price inflation mean that each generation mines the waste dumps of the previous generation. This has been the case in Cornwall, UK, and Laurion, Greece,[3] for over 2,000 years.

The Bingham Canyon copper pit in Utah is the largest man made excavation on Earth and it has been designated a National Historic Landmark. The deposit has produced some 16 million tons of copper and some 9 cubic kilometers of rock have had to be moved to produce this. The ore grade is 0.5% copper and the ore itself constitutes only 2% of the total rock removed: the rest is barren rock that has been moved so that the ore can be accessed. The ore is ground into a very fine powder in order to extract the copper. This fine powder, known as slimes or tailings, is a particularly noxious material that is also extremely reactive.

The tailings are captured in artificial lakes or tailings ponds. These can be very large. The biggest I saw was by the Climax mine in the Rocky Mountains (Figure 7.5). The tailings pond stretched to the horizon with mountain peaks jutting out of it. A memorial tablet records that three towns are buried under it together with the highest Masonic lodge in the United States. The EPA officer I was with when I visited the mine in the 1970s just groaned when I asked him what the EPA was going to do about it. Since that time, the mine has made major advances in land reclamation in the tailings dam areas. For example, tailings are being covered with 2

FIGURE 7.5. The tailings dam at Climax, Colorado. The Rocky Mountains emerge from the dam, which has buried three towns.

feet of rock waste followed by a mulch of biosolids from composted municipal waste and then planted with a native seed mix.

The problem with many of these sites is that the mine itself has closed and the company responsible does not exist anymore. This leaves the clean-up costs to government agencies such as the EPA. The scale of the problem is enormous. There may be as many as 500,000 abandoned mine sites on federal land in the United States alone.[4] If only 5% of these are potential point sources of pollutants, it means that there are 25,000 dangerous sites in need of remediation.

Pyrite is not usually the primary target in the extraction processes in mining so it is relatively concentrated in mine waste. This waste consists of both crushed rock and the chemically treated residue, called tailings or slimes, from the extraction process. These tailings are commonly talcum-powder–sized particles of rock and waste sulfide minerals, such as pyrite, in a slurry with water. These are commonly allowed to settle in dammed lakes and the water is gradually decanted off. This water is a toxic, acidic solution. Because of the oxidation of the fine-grained pyrite, it contains relatively high concentrations of metals—arsenic and selenium, for example—often admixed with a number of complex organic chemicals used in the treatment process. Ultimately this water must be discharged into the environment and today this is done after rigorous treatment to make it safe. Of course, the toxic constituents may be separated from the water but they still have to be disposed of somehow. Before the middle of the 20th century, none of this was done and mine waste was simply pumped into the nearest river, lake, or sea.

The situation is particularly acute with many gold mines. Here the gold, which is the target of the operation, is present usually in parts per million concentrations: in other words you have to treat a ton of ore to get a handful of grams of gold out of it. The other 999,990 g of ore are thrown away. Gold ores are commonly associated with pyrite and arseno-pyrite (the arsenic equivalent of pyrite) minerals, which are dumped in the finely ground tailings and waste rock.

The mine tailings are a combination of a fine, pyrite-enriched powder enriched in toxic metals that is stored in a large lake—it makes a potent mixture. The containment of the tailings in the dam becomes progressively more difficult as the tailings pond becomes bigger. If a bank collapses as in the Aznalcóllar incident in the Rio Tinto district in 1998, the result is the immediate pollution of hundreds of kilometers of the river system and adjacent farmland or nature reserve. Of course when mining ceases the maintenance of the slimes dam also ceases and the poisonous mixture is left to the effects of nature. One of the most serious incidents was the failure of the tailings dam wall at Stava near Trento in Italy in 1985. Some 180,000 cubic meters of tailings poured out of the dam, killing 268 people. The tailings were spread in a layer up to 40 cm deep over an area of more than 4 km². A similar event in Kolontar, Hungary, in 2010 killed ten people and deposited red toxic sludge over 8 km². There are around 3,500 tailings dams in the world and, at present, they are failing at a rate of about two per year.[5] As climate change continues to accelerate and extreme meteorological events become more common, I expect that this rate will increase.

Most people know that coal contains sulfur. What they are often not aware of is that much of the sulfur in coal is in the form of pyrite. The pyrite was formed by sulfate-reducing bacteria in the fetid swamps in which the coal forests grew. As the trees died and decayed they produced an ideal environment for bacterial sulfate-reduction. Aerobic decay of the plant matter, catalyzed by bacteria, used up all the oxygen and produced anoxic conditions. The decay also produced organic nutrients ideally suited to the growth of sulfate reducers. The limiting component was the supply of sulfate. The coal swamps were terrestrial freshwater systems and sulfate was usually less concentrated than in seawater. Marine incursions over coastal swamps occasionally produced a hike in sulfate, which the sulfate-reducing bacteria were ready to exploit. River waters leached sulfate from the rocks and, occasionally, even ran through sulfate-rich rocks, increasing the solution sulfate concentrations.

The consequence of mining coal is therefore to expose pyrite to oxidation. Even though the concentration of pyrite in coal is low relative to the massive sulfide ores discussed previously, the sheer volume of coal mined is enormous. This means that coal mining produces large volumes of acid mine drainage and this is a major threat to rivers, lakes, and groundwaters in coal mining areas. The scale of the problem can be illustrated with respect to the coal mining districts of the Appalachian Mountains in the United States, where some 16,000 km of streams and rivers were estimated to be affected by acid mine drainage in the 1980s.[6] In South Wales, a much smaller area (100 km) of our river systems was polluted by acid mine drainage.

World coal production has doubled from 4 billion tons a year in 1980 to almost 8 billion tons per year in 2011. Much of this increase is due to China, which now accounts for almost half of world coal production. Currently over 30 million tons of sulfur are emitted into the atmosphere annually through coal burning.[7] I estimate that this currently produces about 100 million tons of sulfuric acid per year.[8] This is double the figure in 1980, when most acid production was through coal burning in Europe and the United States, which have subsequently reduced power station sulfur emissions.

There is not much that can be done about acid mine drainage. The remediation techniques include the addition of lime, which precipitates the metals and neutralizes the acid. Alternatively, the acid mine drainage can be directed through constructed wetlands where sulfate-reducing bacteria turn the dissolved metals back into solid sulfides. The problem is: what then? What do we do with the thousands of tons of highly toxic lime waste or soil that the remediation techniques produce? One solution is to put it back in the mine or open pit, but it must be stored in an environment without any air or water access or the same thing will happen all over again. A popular remedy in the United States is solidification and stabilization of the contaminated material by mixing it with a lime-based cement.

Eternal Fires

Early explorers in Australia saw a volcano smoking quietly some 200 km north of Sydney. The native Australians imagined that it was the fiery tears of a woman changed to stone by Biami, the sky god long ago. It turns out to be a coal seam that has been burning for more than 5,500 years.

I mentioned two cases previously—United Verde and San Dionysio—where the massive pyrite deposits were so rich that, when they were exposed to air by the beginning of mining, they ignited spontaneously. This is bad enough, but mining actually continued through the United Verde fire. By contrast, mining coal is far more dangerous from the point of view of combustion: after all, coal is mined as a fuel. While safely buried in the ground and encased in its coal coffin, pyrite is quite inert and does not react. However, as soon as the coal is mined, air gains access and the pyrite oxidizes. As mentioned before, this reaction is exothermic and gives out heat. This is not a good thing in the coal environment. Coal itself burns and many of the gases associated with coal seams, such as methane, can burn explosively. The result can be that the mine catches fire.

If this happens there is not much that can be done about this either. The fire at the Centralia mine in Pennsylvania has been burning steadily underground since 1962. And this itself may be a new outbreak of the nearby Bast colliery fire, which started in 1932. The coal seam steadily burns underground, producing a cocktail of foul-smelling and poisonous gases that can seep to the surface, rendering the area uninhabitable. The surface is baked and large surface collapses have occurred. The fire led to the abandonment of the town of Centralia and, as the underground fire has continued, to the abandonment of the nearby town of Byrnesville. It is anticipated that the fire may burn for another 250 years along a 15-km-long seam, possibly rendering some 15 km^2 of the surface uninhabitable.[9]

The Centralia fire is just one of thousands of coal mine fires that are currently burning worldwide. In 2005, Pennsylvania alone had 38.[10] The Glenwood Springs coal mine fire in Colorado has been burning for over a century. In 2004 it ignited a forest fire that destroyed 50 km^2 and cost $6,500,000 to extinguish. And the mine fire still burns. Some estimates have suggested that uncontrolled fires in 56 of China's 30,000 coal mines contribute several percent of global carbon dioxide emissions. In addition, seams of coal have been burning for centuries in the Ningxia Hui and Xinjiang Uygar autonomous regions in northern China. India currently accounts for the greatest concentrations of coal mine fires and widespread areas of the densely populated coalfields have been turned into wastelands. Coal has been documented to have been burning in the Jharia coalfield in Jharkhand state, eastern India, for example, since 1916 and by the 1960s fires had spread throughout the coalfield.[11] Currently there are more than 70 coal mine fires in Jharia. According to the state government the whole town of Jharia is to be relocated because of the coal fires.[12] There is another reason: the total cost of moving the town and its population is estimated to be $4 billion but there are $12 billion of extractable coal beneath the present town.

The only cure for a coal mine fire is to seal the mine so that air is excluded. The problem is that the hot burning coal produces gases that rise up to the surface and air tends to be sucked into the system to replace these gases. This continues to feed the fire. Mines can be flooded with water or with nitrogen or carbon dioxide gas. Workings above the water table may continue to burn and long-term gas treatment may be expensive and not entirely effective.

Many of the fires associated with coal mining are caused by the waste tips or spoil heaps from the mine catching fire. The cause of the fire is often the same: rapid pyrite oxidation on exposure to air. As with metal mining described above, coal mining produces more waste than coal and the waste tips themselves can contain substantial concentrations of coal. Coal tips burned in South Wales for over one hundred years. The world record was supposedly the waste tip by the South Wales town of Merthyr Tydfil that started burning in the 19th century and was still burning in 1983. The disaster at Aberfan in South Wales, when a mine waste tip buried a school in 1966 and killed 116 children and 28 adults, finally put an end to the *laissez faire* approach to coal mine waste tips. The burning tips were dug up and carted off, the fire was extinguished, and the waste was reburied in landfill. This was helped by the economics of mining. In the old days in South Wales, the inspector at the minehead, who checked the loads of coal the miners were bringing up from underground, would only accept lumps of coal bigger than a man's fist. So all the smaller lumps were thrown on the waste tip, where they would ignite and burn more readily. One Welsh town council were taken aback in 1985 when a man called in to say that he would not only remove the unsightly old coal spoil heap that blighted the town but would do it for nothing. He separated the coal out and sold it. He ended up with a million-dollar profit even after expenses. Since that time, town and city councils have wised up. But it is another example of each generation mining the previous generation's waste.

Coal catches fire by a number of processes, including forest fires and vandalism. Commonly, however, the coal ignites through pyrite oxidation.

Mine waste tips commonly catch fire and are as dangerous as coal mine fires. The difference is that the coal mine fires are more difficult to extinguish. These coal tip fires, on the other hand, can be dug up and removed economically—it just needs a willing local or regional authority.

Acid Rain

An entertaining way to spend an evening is to look at historic headlines in the national press concerning the environment. Environmental concerns go in and out of fashion. In the 1980s the big thing was acid rain. Thus the headline in the UK *Guardian* newspaper almost exactly twenty-nine years ago to the day I am writing this was "Acid Rain Must Be Tackled Now, Say MPs." The story tells us that a committee of Members of Parliament called upon the UK Central Electricity Generating Board (CEGB) to take steps to cut its output of chemicals that cause acid rain. The committee reported that the CEGB was the main producer of sulfur dioxide, SO_2, which is a major cause of acid rain. The sulfur dioxide is produced by burning the pyrite in coal in power stations. This gas dissolves in acid water to fall as acid rain, a dilute solution of sulfuric acid. It seems astonishing to us today that the *Guardian* reported that the CEGB denied that coal burning had any relation to acid-rain formation. Prime Minister Margaret Thatcher privatized the CEGB shortly after this.

Burning pyrite produces SO_2, which dissolves to form acid. The phenomenon has a long history. The diarist John Evelyn noted the effects of city air on limestone and marble in 17th-century London.[13] Robert Angus Smith, a Scottish chemist, first demonstrated the relationship between acid rain and atmospheric pollution in 1852. Smith coined the term *acid rain* in 1872.[14]

Acid rain affected eastern Europe, Scandinavia, the eastern United States, southeast Canada, and southeast China particularly. It has had a major effect on forests, and the sight of swathes of bare-branched, leafless, and dying conifers in Scandinavia and even in Wales was dramatic in the 1980s. Acid rain also caused acidification of inland watercourses, and, as the water became acid, the biodiversity of lakes and rivers decreased. The acidification was not as intense as with acid mine drainage, but acid rain causes damage on a regional scale. For townies and tourists, one of the most dramatic effects was the corrosion of limestone buildings and the defacement of many of the statues and carvings.

In 1980 the US Congress initiated a ten-year research program into acid rain. The resulting report stated in 1991 that 5% of New England lakes were acidic. In 1990 Congress passed a series of substantive amendments to the Clean Air Act controlling emissions from US power plants. The result has been that US SO_2 emissions have been cut by 40%.[15] During the same period, SO_2 emissions from power plants in the European Union dropped by 70%.[16]

The huge increase in coal burning in China has had a significant effect, and currently almost one-third of China is experiencing acid rain. The main amelioration technology is readily available and has been widely applied. These are scrubbers on the power station chimneys that trap the SO_2 and fix it as calcium sulfate. Calcium sulfate is a major component of cement, and therefore much of this can be recycled. At present around 30% is recycled in the United States and 70% buried in landfills. Using scrubbers adds around only 10% to electricity production costs so that removal of pyrite-sourced SO_2 from power-station coal is economically accessible and technically efficient. It has been mainly responsible for the decrease in European and North American emissions. It has also been responsible for a decrease in sulfur dioxide emissions in China in the past five years even though the amount of coal being burned in power stations has continued to increase.

Smog

Returning to our survey of newspaper environment headlines and how they have changed with time, we now look at the UK newspapers in the 1950s. The headline problem in London at that time was *smog*, a portmanteau word referring to a mixture of smoke and fog. The mixture includes soot particles and sulfur dioxide. Both are derived from coal burning: the soot from the coal and the sulfur dioxide from the pyrite. The modern fogs in cities such as Denver and Phoenix are mainly caused by vehicular emissions, which react with sunlight to form a photochemical smog. Pyrite is not directly involved in the formation of photochemical smogs.

Smogs have been around for centuries. Edward I banned coal fires in London in 1306 in response to increasing air pollution. The London smogs were a characteristic of the city for several hundred years, and several attempts were made by various monarchs to ban coal burning in the city. The smogs were primarily caused by the burning of coal in houses for heating and cooking and were a feature of cities and towns, areas where many homes were crowded together. The industrial contribution varied from district to district. In 1661 John Evelyn was put up to writing a polemic against coal burning in London by Charles II:[17]

> But that Hellish and dismal Cloud... So universally mixed with that otherwise wholesome and excellent Aer, that her Inhabitants breathe nothing but an impure and thick Mist, accompanied with a fuliginous and filthy vapour, which renders them obnoxious to a thousand inconveniences, corrupting the Lungs, and disordering the entire habit of their Bodies; so that Catharrs, Phthisicks, Coughs and Consumptions, rage more in this one City, than in the whole Earth beside.

We used coal as the primary heating source in our house during my childhood. Each fall the coal man would come and load around a ton of coal in hundredweight bags into the coal shed. Translating this, about 1,800 kilograms of coal would be delivered each fall in 50-kilogram hessian sacks and piled up in a shed dedicated to this purpose. Some of the older houses had coal cellars. In this case the coal was tipped through a manhole in the sidewalk to a cellar beneath the house. The coal man always wore a flat cap and ported the coal sacks on a flatbed wagon pulled by a blinkered horse. The coal he delivered would last us for heating through to the summer. It was good bituminous coal in large lumps often with pyrite glittering in the black mass.

The same process was repeated in millions of homes throughout the country. Each ton of coal burned produced perhaps 1 kilogram of sulfur dioxide mainly from the pyrite.

Smogs still feature heavily in period dramas, since they add to the atmosphere of mystery. T.S. Eliot described them best in the *Love Song of J. Alfred Prufrock*:

> *The yellow fog that rubs its back upon the window-panes,*   15
> *The yellow smoke that rubs its muzzle on the window-panes*
> *Licked its tongue into the corners of the evening,*
> *Lingered upon the pools that stand in drains,*
> *Let fall upon its back the soot that falls from chimneys,*
> *Slipped by the terrace, made a sudden leap,*   20

And seeing that it was a soft October night,
Curled once about the house, and fell asleep.

And indeed there will be time
For the yellow smoke that slides along the street,
Rubbing its back upon the window panes;

The smog problem in the United Kingdom culminated in the great London pea-souper of 1952. A pea-souper was so called because of the yellow-green color of the smog—which I assume was caused by the sulfur content produced from the burning pyrite—and the reduction of visibility to less than 1 meter—the "yellow smoke" referred to by Eliot. My mother was caught in the 1952 London smog, and she said that visibility was so bad she could not see her hand in front of her face. Over 4,000 people died from bronchitis, asthma, and emphysema as a direct result of the 1952 smog and a further 8,000 died in the following weeks. The government rapidly introduced the Clean Air Act, which banned the burning of coal in many towns. The result was the disappearance of the classic London smog. It also enabled a clean-up of the city buildings as the soot and grime were scraped off and did not return.

A similar approach was taken by the Beijing authorities in the preparation for the Beijing Olympics in 2008. The result was a dramatic improvement in the air quality compared with earlier years. However, I still experienced smogs in Beijing in 2008 and a major smog event occurred in 2011. Coal is still being burned in large quantities for heating and the city is affected by coal burning in adjacent regions. The problem now is that the exponential increase in the use of cars in Beijing has caused the increase in photochemical smogs, which exacerbate the problem. The authorities have capped new automobile permits in the city in order to start to combat this problem. Even so, 2013 was perhaps the worst year for air pollution in Beijing and many other cities of northern China.

Acid Sulfate Soils

Acid sulfate soils have been described as the nastiest soils on Earth. They are fundamentally natural phenomena caused by the oxidation of soils containing pyrite. The oxidation produces sulfuric acid that causes adverse environmental, ecological, health, and economic effects. The acidification of the soils leads to a deficiency in essential plant nutrients and plant base minerals such as calcium, magnesium, and potassium. At the same time, potentially toxic metals such as aluminum, iron, manganese, and other heavy metals increase. The concentration of nitrogen-fixing micro-organisms in the soil can also be negatively affected. The net effects are a weakening of plant resistance to pathogens and poor plant productivity. Waterway habitats are degraded as food resources are destroyed and the environment becomes more acidic. The effects on human health are a subject of current worldwide research. High aluminum concentrations in drinking and bathing waters, for example, are already known to cause stunted growth and mental impairment to exposed populations.

Linnaeus first described acid sulfate soils in 1735 on a visit to Holland. He called them *argilla vitriolacea* using his typical binomial system that he applied to the classification of plants. *Argilla vitriolacea* simply means clay mixed with sulfuric acid. The problem of acid sulfate soil in the polders of Holland and the long-term Dutch experience of acid sulfate soils provided a major resource for identification and remediation worldwide. I attended the first international symposium on acid sulfate soils, which was appropriately held at Wageningen in the Netherlands in 1972. I was there to present a keynote paper on how the pyrite got into the soil in the first place: the waterlogged soils had become anoxic, sulfate-reducing bacteria then used the organic matter in the soil to reduce sulfate in the water to sulfide, and this reacted with soil iron minerals to form pyrite.[18]

Acid sulfate soils occur primarily in coastal areas, or former coastal areas, since the source (pyrite) is formed by the reduction of sulfate, which is more concentrated in seawater. Waterlogged pyrite-containing soils are relatively safe since air is generally excluded from contact with the pyrite. But as soon as the soil is raised above the water table, the pyrite is exposed to air with the now-familiar effect of acid production. The usual suite of yellow and red iron oxides can be precipitated, and these can be complemented by iron sulfate minerals such as jarosite. Soils mottled with straw-yellow jarosites are often called *cat clays*. As in the case of cat gold in Chapter 1, the *cat* here refers here to the mysterious, harmful—literally heretical—properties of the clays. It was originally used colloquially in Holland to describe sulfidic muds and other varieties of infertile soils, including mottled varieties. The use of cat clay to describe acid sulfate soils specifically stems from the first regional soil investigations in Holland in 1923. The formation of acid sulfate soils in the Dutch polders came as a complete surprise to the engineers originally responsible for draining the land. The waterlogged sediments did not look much different than other soils that had been successfully turned into highly productive areas by drainage.

The acid attacks the clay minerals in the soil to release aluminum. The situation can be exacerbated by climate change, drainage schemes, and elevation of the land. Although acidification of soils is a natural phenomenon, it has become apparent that human-induced soil acidification is currently the dominant factor in the global acidification of soils. The importance of this to the global surface environment is apparent when it became understood that the soil was a key component of the Earth's "critical zone": the heterogeneous, near-surface environment of the planet in which complex interactions involving rock, soil, water, air, and living organisms regulate the natural habitat and determine the availability of life-sustaining resources. The soil is the membrane through which chemicals and life forms interact with the hydrosphere, atmosphere, and geosphere to create a life-sustaining surface on the planet. In the beginning of this millennium it was realized that we really have little idea about the biology, chemistry, and geology of this critical zone for life on Earth. The first widespread use of the concept was by the US National Research Council in its 2001 report on basic research opportunities in the Earth sciences, a sort of vademecum for US researchers indicating the areas they were likely to get funding for in the medium term.[19] The first international meeting to promote a worldwide effort of interdisciplinary and multidisciplinary approaches to critical zone research was held at the University of Delaware in 2005.

Acid sulfate soils are a major problem in Australia, where some 2 billion tons of pyrite in coastal soils could potentially oxidize to produce 3 billion tons of sulfuric acid.[20] In Finland and Sweden the land is rising relatively rapidly as a result of the melting of the ice from the last Ice Age, some 8,500 years ago. The rate in the Stockholm area is around 1 meter per 100 years, and this is why medieval ships, such as the *Vasa*, keep appearing in Stockholm harbor. It also means that pyritic sediments are being exposed to the atmosphere and oxidizing to form acid sulfate soils. Over 170,000 square kilometers of acid sulfate soils have been identified worldwide, with 65,000 in Asia; 45,000 in Africa; 30,000 in Australia; 30,000 in South America; and some 10,000 in Europe and North America.[21]

The successful reclamation of acid sulfate was originally achieved by Dutch farmers who simply plowed to a depth of greater than 30 cm so that lime from the underlying bedrock was brought up to the surface and neutralized the acid. This led to the basic remedy for acid sulfate soils: moving large tonnages of earth and the addition of lime to the soils in sufficient quantity to neutralize the acid. This is expensive. It may be possible in limited areas of acid sulfate soils in the rich countries of western Europe, North America, and Australia, but it may not be feasible over larger areas in developing countries.

Acidogene

The oxidation of pyrite is a natural reaction that, as is shown in Chapter 8, is a fundamental part of the global biogeochemical cycles. These global cycles have buffered the Earth's surface environment, making it suitable for life for a greater part of its 4.5-billion-year history. This buffering effect is still working with the current dominant anthropogenic or humanmade acid production. For example, the burning of pyrite in coal in power stations is a major factor in the production of acid rain. However, the sulfate that is spewed into the atmosphere forms aerosols, tiny droplets of sulfuric acid and sulfate particles that scatter and reflect sunlight and also increase the reflectivity of the Earth, since they serve as cloud condensation nuclei. Increased numbers of condensation nuclei increase the density of clouds and increase cloud water content. Sulfate aerosols are thus effective coolants for the Earth. Indeed, increased power-station coal burning in the latter decades of the 20th century has been proposed as a major reason why carbon dioxide emissions have not produced the expected global temperature rises. President George W. Bush said, with characteristic insight, that in order to counteract global warming, people should simply burn more coal. The effect of potential reduction in sulfur dioxide emissions from burning pyrite in coal-fueled power stations is that the rate of global warming may increase. Climate scientists have proposed increasing sulfate aerosol concentrations by artificially injecting sulfate into the stratosphere with balloons or rockets, thus combating these anticipated temperature rises.[22]

The injection of sulfate aerosols into the upper reaches of the atmosphere is also a natural process. It occurs during explosive volcanic eruptions where the power is so great that sulfur dioxide can be thrown into the stratosphere. However, most of the haze we see in industrialized countries is produced through power stations burning pyrite in coal. I must confess that I had not realized that pyrite burning was a major cause of regional atmospheric haze when I started this book. I thought that the veil of haze that blankets the central and eastern United States, often for days on end, was a natural phenomenon. I was put right by Stephen Corfidi of the National Weather Service of the US National Oceanic and Atmospheric Administration whose original article in the *National Weather Digest* of 1996 has regularly been updated on the Web through 2013.[23] I am old enough to remember that the characteristic haze over the Blue Mountains and Smoky Mountains of the eastern United States and the Blue Mountains of Australia was originally related to emissions from plants. Of course a small fraction is still derived from evapotranspiration of plants, but even in these iconic areas, haze today is mainly caused by pyrite burning. These regional hazes dim the sun and put an early end to sunlit days. The hazes are formed by exactly the same processes that caused the smogs described earlier. The pyrite burning produces sulfur dioxide, which dissolves in atmospheric water to form sulfuric acid droplets. These droplets in themselves are potentially injurious to human health, especially to lung tissues and breathing passages. The haze enhances the destructive nature of the photochemical smogs. Thus although the photochemical smogs are not caused directly by pyrite burning, pyrite burning does contribute to the deleterious effects of this phenomenon. The haze also contributes to reducing the amount of solar energy reaching the Earth's surface. The estimate for the United States is a decrease of about 7.5%.[24] Even though the incidence of these pyrite-induced hazes is decreasing in the United States, there is some evidence that haze is becoming a global phenomenon. Hazes generated in southeast Asia and China appear to be beginning to be detected in the United States.

Much of this discussion about pyrite burning has concentrated on continental effects such as acid rain, acid mine waters, and smogs. However, the continents cover only about 30% of the Earth's surface area; the rest is ocean. The acidification of the land by pyrite burning may therefore be thought to represent only a minor component of the Earth system. The dramatic increase in anthropogenic CO_2 emissions over the past century has resulted in a steady acidification of the oceans. Thus at the same time that land systems are becoming more acidic through pyrite burning, the oceans are also becoming more acidic. This significantly affects the acid budget of the Earth. Normally, seawater is slightly alkaline and the vast volumes of the oceans are sufficient to neutralize the acidic run-off from the continents. However, today the oceans are becoming increasingly more acidic so that this neutralization capability is being attenuated.

The net effect of these trends appears to be that the surface of the Earth is becoming more acidic. The late US ecologist Eugene Stoermer originally suggested calling the era we are living in the *Anthropocene* to describe the overwhelming influence of human activities on the planet, and this was popularized by the Nobel Laureate Paul Crutzen. From a geochemical point of view it might be called the *Acidogene*, since future geochemists will see little evidence of human existence but the rocks will preserve the characteristic signatures of global acidification. These will include a red layer in which much of the metals have been leached out by acid, the major form of iron is ferric rather than ferrous, and pyrite is only conspicuous by its absence. The fossil remains will show major migrations as animals move away from acidified regions of the land and oceans, which no longer support their life forms. Ultimately, the fossil record will show a mass extinction event, which will be interestingly characterized by the demise of many calcium-carbonate–shelled creatures as they dissolve more rapidly in acidic oceans. Of course, the layer will be very thin. As pointed out by the US geochemist John Morse, the Earth system will ultimately remediate the acidification of the oceans. It will probably take about 1,000 years. In the meantime, of course, about thirty generations of humans will have to survive it. James Lovelock, one of the originators of the Gaia hypothesis, dryly observed that life on Earth will survive the exigencies of these temporary environmental aberrations caused by our species. The Earth environment will return to balance in time. The only worry is whether the human race will still be part of it. And what will be the state of *Homo sapiens* then?

Notes

1. See D. Rickard and G.W. Luther. 1997. Kinetics of pyrite formation by the H_2S oxidation of iron(II) monosulfide in aqueous solutions between 25 °C and 125 °C: the mechanism. *Geochimica et Cosmochimica Acta*, 61:135–147. Here we showed that H_2S could act as quite a decent oxidizing agent, a concept that is so revolutionary that it still echoes around some of the ivory towers of Academe today.

2. See L.U. Slakield. 1987. *A Technical History of the Rio Tinto Mines from Pre-Phoenician times to the 1950s* (Houten, The Netherlands: Springer), 500pp.

3. Laurion is the site of the silver mines, which kept classical Athenians rich enough to employ slaves while they wandered around the groves of Academe. In 483 BCE Thermistocles persuaded the Athenians to invest these riches into building a navy. The result was the victory of the Athenians over the Persians at the battle of Salamis three years later, which changed the course of European

civilization. A French company was still working the dumps at Laurion in the 1970s.

4. Abandoned Mine Lands Portal: www.abandonedmines.gov/ep.html.

5. M.P. Davies and T.E. Martin. 2000. Upstream constructed tailings dams—A review of the basics. In *Proceedings of Tailings and Mine Waste '00* (Fort Collins, CO: Balkema), pp. 3–15.

6. A.T. Herlihy, P.R. Kaufman, M.E. Mitch, and D.D. Brown. 1990. Regional estimates of acid mine drainage impact on streams in the mid-Atlantic and southeastern United States. *Water, Air and Soil Pollution*, 50:91–107.

7. R.K. Kaufmann, H. Kauppi, M.L. Mann, and J.H. Stock. 2011. Reconciling anthropogenic climate change with observed temperature 1998–2008. *Proceedings of the National Academy of Sciences of the United States of America*, 108:11791–11794.

8. One mole S gives 1 mole H_2SO_4. Note my guestimate assumes 1% pyrite in average coal. Four moles of acid from 2 moles of pyrite gives the same result.

9. K. Krajick. 2005. Fire in the hole. *Smithsonian Magazine*, p. 54. Retrieved October 24, 2006.

10. See the *US Office of Surface Mining Reclamation and Enforcement (OSM), Abandoned Mine Land Inventory System (AMLIS)*. They note that their figures are probable underestimates.

11. G.B. Stracher. 2002. Coal fires: a burning global recipe for catastrophe. *Geotimes*, 47:36–43.

12. *The Times of India*, August 31, 1996 (accessed December 8, 2013).

13. E.S. de Beer, ed. 1955. *The Diary of John Evelyn*, Vol. 3. September 19, 1667 (Oxford: Clarendon Press), p. 495.

14. R.A. Smith. 1852. On the air and rain in Manchester. *Manchester Literary and Philosophical Society Memo*, 10:207–217; 1872. *Air and Rain: The Beginnings of Chemical Climatology* (London: Longmans, Green and Co.), 600pp.

15. See, for example, D.A. Burns, ed. 2011. *National Acid Precipitation Assessment Program Report to Congress 2011: An Integrated Assessment* (Washington, DC: Executive Office of the President of the United States), 132pp. This is an updated review of the progress with acid rain and haze in the United States and an excellent introduction to the problem.

16. T. Gilberston and O. Reyes. 2009. Carbon trading: How it works and why it fails. Critical Currents Occasional Paper 7 (Uppsala: Dag Hammarskjöld Foundation), 102pp.

17. J. Evelyn. 1661. *Fumifugium: or, The Inconvenience of the Aer and Smoake of London Dissipated, Together with some Remedies humbly proposed by J. E. Esq., to His Sacred Majestie, and to the Parliament now Assembled* (London: Gabriel Bedel and Thomas Collins). Reprinted in 1999. *Organization and Environment*, 12:187–193.

18. D. Rickard. 1972. Sedimentary iron sulphide formation. In *Proceedings of the International Symposium on Acid Sulphate Soils* (Wageningen, The Netherlands: International Institute for Land Reclamation and Improvement), pp. 28–65.

19. Committee on Basic Research Opportunities in the Earth Sciences. 2001. The critical zone: Earth's near-surface environment. In *Basic Research Opportunities in Earth Science* (Washington, DC: National Academy Press), pp. 35–45.

20. R.W. Fitzpatrick, B. Powell, and S. Marvanek. 2008. Atlas of Australian acid sulfate soils. In *Inland Acid Sulfate Soil Systems Across Australia*, edited by R.W. Fitzpatrick and P. Shand. CRC LEME Open File Report No. 249. Thematic Volume (Perth, Australia: CRC LEME), 90pp.

21. W. Andriesse and M.E.F. van Mensvoort. 2006. Acid sulfate soils: distribution and extent. In *Encyclopaedia of Soil Science*, edited by R. Lal (Boca Raton, FL: CRC).

22. For example, T.M.L. Wigley. 2006. A combined mitigation/geoengineering approach to climate stabilization. *Science*, 314:452–454.

23. S.F. Corfidi. 1996. Haze over the central and eastern United States. *National Weather Digest*, March.

24. R.J. Ball and G. D. Robinson. 1982. The origin of haze in the central United States and its effect on solar radiation. *Journal of Applied Meteorology*, 21:171–188.

FIGURE 3.1. Massive pyrite: part of a wall of golden massive pyrite several meters high from Udden, northern Sweden.

FIGURE 3.2. Typical Victorian marcasite brooch showing polished and faceted pyrites set in sterling silver. Courtesy of www.antique revisions.com.

FIGURE 3.3. Pyrite flower: a broken surface of a 5-cm diameter pyrite nodule showing the radiating pyrite crystals.

FIGURE 4.1. Pyritized ammonite from Charmouth, England. Photo M. Keating.

FIGURE 4.9. Pyrite cubes from Navajún, Spain. These 5 cm crystals are in their natural state: they have not been cut or polished.

FIGURE 4.10. Striated pyrite cubes from Huanzala, Mexico. Photo James Murowchick.

FIGURE 4.12. Pyritohedra. Although these are the nearest forms to the dodecahedron, inspection of the large crystal on the right, for example, shows that the faces are not the same size nor are they regular pentagons.

FIGURE 4.14. Pyrite octahedra from the Huazala Mine, Huallanca District, Huanuco Department, Peru. Photo Carlos Millan.

FIGURE 5.2. Looking into the mouth of Hell. The Hekla eruption of 1980. The lava is at a temperature of around 1100°C, which compares with a hot oven at 200°C and molten lead at 327°C.

FIGURE 5.5. The effect of 300 years oxidation in a massive pyrite ore. A collapsed stope in the Falun mine, Sweden, showing iron (green), copper (turquoise), and calcium, magnesium, zinc, and lead (white) sulfate stalactites in a mass of iron oxide ocher. The temperature can reach 50°C here. Photo Folke Schippel and Rolf Hallberg. From D. Rickard and R.O. Hallberg. 1973. De små gruvarbetarna i Falun. *Svensk Naturvetenskap*, 102–107.

FIGURE 5.7. Sulfur-rich emanations, including pyrite as well as rarer minerals such as orpiment and gold, in the boiling geothermal springs of Champagne Pool, Waiotapu, New Zealand.

FIGURE 5.14. Pyrite-rich sediments from the c. 20 million year old, Kuroko district, Japan. The pyrite is forming classical sedimentary structures and shows graded bedding with coarser grains toward the base. The 20th-century debate was whether these textures showed that pyrite was formed in volcanic sediments or hot, pyrite-bearing solutions had penetrated the rocks later, mimicking the earlier sedimentary structures. Photo William Sacco courtesy of the Society of Economic Geologists. From H. Ohmoto and B.J. Skinner, eds., 1983. *The kuroko and related volcanogenic massive sulfide deposits. Economic Geology Monograph,* 5: cover image.

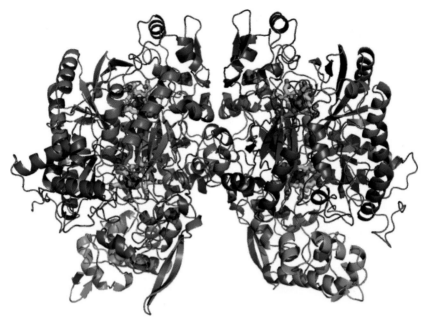

FIGURE 9.1. The structure of dissimilatory sulfite reductase, a key enzyme in biological sulfate reduction. The different colors show different types of atoms: yellow carbon, red oxygen, blue nitrogen, brown iron, and green sulfur. From T.F. Oliveira et al. 2008. The crystal structure of *Desulfovibrio vulgaris* dissimilatory sulfite reductase bound to DsrC provides novel insights into the mechanism of sulfate respiration. *Journal of Biological Chemistry*, 283:34141–34149.

FIGURE 9.3. A. Modern stromatolites in the Hamelin Pool, Western Australia. B. Similar 3.47 billion-year-old stromatolites from Western Australia. From D. Rickard. 2012. *Sulfidic sediments and sedimentary rocks* (Amsterdam: Elsevier, 801pp.). Reprinted by permission.

FIGURE 10.1. Hidden gold. Small blebs of yellow gold (arrowed) within a 0.3 mm pyrite crystal. The photomicrograph was taken with a blue filter to show the yellow gold more clearly against the normally bright, yellow-white of the pyrite grain.

FIGURE 10.2. A happy prospector with a handful of gold.

8

Pyrite and the Global Environment

THE TWO BASIC processes concerning pyrite in the environment are the formation of pyrite, which usually involves reduction of sulfate to sulfide, and the destruction of pyrite, which usually involves oxidation of sulfide to sulfate. On an ideal planet these two processes might be exactly balanced. But pyrite is buried in sediments sometimes for hundreds of millions of years, and the sulfur in this buried pyrite is removed from the system, so the balance is disturbed. The lack of balance between sulfide oxidation and sulfate reduction powers a global dynamic cycle for sulfur. This would be complex enough if this were the whole story. However, as we have seen, both the reduction and oxidation arms of the global cycle are essentially biological—specifically microbiological—processes. This means that there is an intrinsic link between the sulfur cycle and life on Earth. In this chapter, we examine the central role that pyrite plays, and has played, in determining the surface environment of the planet. In doing so we reveal how pyrite, the humble iron sulfide mineral, is a key component of maintaining and developing life on Earth.

Pyrite Formation and Burial

In Chapter 4 we concluded that Mother Nature must be particularly fond of pyrite framboids: a thousand billion of these microscopic raspberry-like spheres are formed in sediments every second. If we translate this into sulfur production, some 60 million tons of sulfur is buried as pyrite in sediments each year. But this is only a fraction of the total amount of sulfide produced every year by sulfate-reducing bacteria. In 1982 the Danish geomicrobiologist Bo Barker Jørgensen discovered that as much as 90% of the sulfide produced by sulfate-reducing bacteria was rapidly reoxidized by sulfur-oxidizing microorganisms.[1] Sulfate-reducing microorganisms actually produce about 300 million tons of sulfur each year, but about 240 million tons is reoxidized.[2]

The magnitude of the sulfide production by sulfate-reducing bacteria can be appreciated by comparison with the sulfur produced by volcanoes. As discussed in Chapter 5, it was previously supposed that all sulfur, and thus pyrite, had a volcanic origin. In fact volcanoes produce just 10 million tons of sulfur each year.[3] This represents only 3% of the total global sulfur. Even adding in the sulfur produced by black smokers from deep-ocean hydrothermal vents adds only 1 million tons of sulfur each year to the total.[4]

About 96% of global sulfide production is through the activities of sulfate-reducing bacteria in sediments, soils, and waters worldwide. The sulfur produced by volcanoes is mostly oxidized and ends up in the oceans or atmosphere. Even if we assume that all 10 million tons of volcanic sulfur produced annually is fixed in pyrite, this is still a small fraction of the 60 million tons of pyrite that is buried in sediments each year. And it is probable that less than 1% of the volcanic sulfur is actually fixed in pyrite.

We therefore end up with the counterintuitive idea that volcanic sulfur is only a small fraction of the total global sulfur and that, for all intents and purposes, global sulfur production can be closely approximated by bacterial sulfate reduction. The ancient idea, held by humankind for millennia, that pyrite is mainly formed in volcanoes is false. It took over 2,000 years of research to show that, in fact, most pyrite is formed by bacteria in sediments.

In contrast to many scientific discoveries, there was no single person responsible for this fundamental advance in our understanding of the natural world. If anyone was first to recognize the importance of sedimentary pyrite to the global sulfur system, it was probably Ken Sugawara and his team from Nagoya University, Japan, in 1953.[5] However, these results were published in a local journal and took some time to be widely circulated. As we have seen in Chapter 6, sulfate-reducing bacteria had been discovered by Beijerinck in the closing years of the 19th century, and the *Challenger* expedition had concluded that reactions of seawater sulfate with organic matter were responsible for the global distribution of blue muds. Even so, the global importance of these microorganisms in the sulfur cycle was not appreciated at that time. By the 1930s it was realized that bacterial sulfate reduction was an important biological process. Indeed in 1934 the successors to Beijerinck at the Delft School of Technology[6] could write, "Sulphate reduction is one of the most commonly occurring and extensive microbiological processes on Earth." It gradually dawned on the scientific community in the latter parts of the 20th century that the myriad results that individual scientists and groups of researchers were collecting from sediments showed that pyrite is mainly formed in sedimentary environments.

Pyrite as a Key Component of the Earth System

Sulfate-reducing bacteria are able to produce this enormous amount of sulfide for two major reasons. First, sulfate is far more abundant in seawater than dissolved oxygen. Even in contact with our present atmosphere, only a few milligrams of oxygen are soluble in a liter of seawater. The number is imprecise because it depends on temperature, salinity, and pressure, but the average dissolved oxygen concentration in the oceans is 3 mg per liter. By contrast, there are around 2,700 mg of sulfate in a liter of seawater. Therefore sulfate-reducing bacteria have almost 1,000 times more of their preferred electron donor than the everyday oxygen breathers.

Second, the number of microorganisms in sediments is astonishing. The average sediment contains at least 1 million cells per cubic centimeter, and this may reach over 10 billion cells per cubic centimeter in particularly agreeable environments. The size of bacteria is not truly appreciated until it is realized that even this vast number of organisms physically represents just 0.005% of the total sediment volume. It is difficult to state with any certainty what proportion of these microorganisms in the sediment are active sulfate reducers, and it will vary between different sites. One problem is that over 90% of all bacteria found in sediments have not been cultured; they are only known from their DNA signatures. Even then, there are probably large numbers of microorganisms that have not been identified by molecular probes since these were only produced in the first place from microorganisms that had been cultured in the laboratory. More recent methods have involved producing molecular probes from just a small DNA sequence, making these probes far more generic in their applicability.[7]

As discussed in Chapter 7, a reduction reaction, such as the reduction of sulfate to sulfide by sulfate-reducing bacteria, must be balanced by an oxidation reaction. This preserves the electrical neutrality of the system. The balancing oxidation reaction for sulfate reducers is the oxidation of organic matter, ultimately forming carbon dioxide. In his classic 1982 study, Bo Barker Jørgensen also measured the relative importance of sulfate-reducing bacteria by analyzing how much of the organic matter supply to sediments they oxidize. He discovered that over half of all the organic matter reaching the sediment is oxidized by sulfate-reducing bacteria with the ultimate production of carbon dioxide. This is an astonishing statistic. It is like saying that over half of the available food on Earth is consumed by humans. The rest of the consumption of organic matter is due to the activities of aerobic organisms like us, which require oxygen to grow.

The sulfate-reducing bacteria are essentially anaerobic; that is, they grow only in anoxic, oxygen-free environments. The statistic regarding organic matter oxidation by sulfate-reducing bacteria underlines the fact that most of the Earth is anoxic. The familiar oxygen-rich environment that we inhabit is just a thin skin on the surface of the planet. The dominant environment of planet Earth is the anoxic environment, which is populated mainly by sulfate-reducing bacteria. Pyrite can be regarded as the leftover from the activities of the sulfate-reducing bacteria. It is a tracer of their presence now and in the past.

The fact that bacterial sulfate reduction alone accounts for at least half of the organic matter supplied to marine sediments provides a strong connect between pyrite formation and biology. Photosynthesis at the sea surface is the driver for the Earth's biosystem. This is called *primary production* since it converts inorganic compounds, including carbon dioxide, into organic matter using sunlight. The organic matter produced by primary production that reaches the seafloor is turned back into carbon dioxide mainly by the activities of sulfate-reducing bacteria. This has enormous consequences for the Earth's biosphere. It means that the amount of carbon buried as organic matter in sediments is basically the sulfate-reducing bacteria's leftovers. The more organic matter that is supplied, the more sulfide the sulfate-reducing bacteria produces and the more pyrite is formed in the sediment. The only way to limit this is to restrict the supply of sulfate. However, the oceans provide an almost infinite sulfate supply for the bacteria. Thus pyrite formation is related to primary productivity, the basic process that drives life on Earth.

The burial of organic carbon can be regarded as removing carbon dioxide from the atmosphere. This is a well-known effect in terrestrial environments, at least in the concerns about global warming, especially anthropogenic warming. The burning of wood, for example, is not regarded as a net contributor to increased carbon dioxide if the wood comes from sustainable forests. This is not because the trees are photosynthesizers that turn carbon dioxide into oxygen, since they only do this in daylight. At night they carry out the reverse reaction, turning oxygen into carbon dioxide, so there is a balance. The reason is that carbon is locked up in organic matter in the trees in the short term and in the long term some of this tree carbon is buried in sediments. Then the oxygen–carbon dioxide balance is skewed in favor of oxygen. In the ocean the process is magnified many times over, since the ocean is the site of primary productivity. Here the burial of organic matter ultimately determines the oxygen content of the atmosphere. Because of the overwhelming importance of sulfate-reducing bacteria to the fate of sedimentary organic matter, the amount of organic matter buried in sediments is largely determined by the activities of sulfate-reducing bacteria. These activities produce pyrite so, ultimately, the amount of oxygen in the atmosphere is a function of the amount of pyrite buried in sediments.

The formation and burial of pyrite is a key component of the oxygen and carbon cycles of the Earth. As such, it is directly related to biological productivity and to the Earth's climate. Since biological processes require other key elements such as nitrogen and phosphorus, pyrite formation helps in determining their availability at any time too.

When pyrite crystallizes in sediments, a wide variety of metals other than iron, including copper, nickel, lead, zinc, and cobalt and other metalloids, such as arsenic and selenium, may be fixed in the pyrite crystals. The amount of these elements fixed in pyrite depends to a large extent on how much was available in the sediment porewaters at the time, and this is related to the concentrations of these elements in the oceans and associated waters. So pyrite formation is also a key component in the global cycles of many metals. The fact that pyrite formation is directly implicated in the global biogeochemical cycles of so many elements means that pyrite is a key component of the Earth system. The nice thing about pyrite is that it can be preserved over geologic time periods, and so we can probe pyrite to obtain information not only about the past environments on Earth but also how these environments evolved. Much of what we know about the history of the Earth has come from investigations of ancient pyrite.

We can also see that pyrite formation is a key component of the present Earth system. The amount of pyrite that is being formed in sediments today should ideally be balanced by the amount of sulfate being produced annually by pyrite oxidation. Otherwise the sulfate concentration of seawater would change. In fact, in 1971 a young Yale geochemist named Bob

Berner showed that the sulfur cycle was not in balance and that the sulfate concentration in the oceans was increasing. He calculated that pollution had increased the river-borne flux of sulfate by about 27%. By 1987 Berner had revised this estimate to 47%.[8] Even though the amount of sulfate in the oceans is enormous, the rate of change is sufficient for it to have a significant effect.[9] Berner went on to provide the basic understanding of biogeochemical cycles in the Earth and showed how the abundance of key elements in the Earth surface environment evolved over geologic time. Pyrite formation formed a cornerstone of these discoveries.

The fundamental role that pyrite plays in the global oxygen and carbon dioxide cycles was first recognized in 1845 by Jacques-Joseph Ebelmen, a distinguished French chemist at the Ecole des Mines in Paris.[10] According to Ebelmen, the formation of pyrite was a process that caused a decrease in the proportion of carbon dioxide in the atmosphere with the liberation of oxygen. He also realized that pyrite formation could raise the proportion of atmospheric oxygen by decreasing the proportion of carbon dioxide. Finally he listed the oxidative weathering of pyrite as a process that would decrease the proportion of oxygen in the air and produce carbon dioxide. These reactions can be summarized in a single, balanced chemical equation,[11] which states that pyrite reacts with oxygen and water to produce iron oxide, sulfate, and acid. This is the familiar oxidation reaction that ultimately supplies sulfate to the oceans. Since the equation is balanced, we can look at it the other way round. It also states that iron oxide reacts with sulfate and acid to produce pyrite and water and release oxygen. This of course is the summary of the process forming pyrite in sediments through the activities of sulfate-reducing bacteria. Since these bacteria are the major metabolizers of the organic matter supplied to sediments, the process of pyrite formation is also related to a separate balanced equation[12] where carbon dioxide and water undergo photosynthesis to produce organic matter and oxygen.

In the latter half of the 20th century researchers realized that these equations, since they were balanced, could be quantified. The advent of computers meant that models of the system could be run and compared with, especially, information on the amount of pyrite buried in sediments through time. The essential role of pyrite in this cannot be overemphasized. The level of atmospheric oxygen at any time is simply the difference between the oxygen flux produced by organic carbon and pyrite burial in sediments and the oxygen removed by the oxidation of organic carbon and pyrite.

The amount of pyrite in sedimentary rocks is a measurable quantity, and this is fundamental to the system. Thus the models could be run and evaluated on how nearly they reproduced the known history of sedimentary pyrite. This was further constrained by measurements of the changes in the sulfur isotope compositions of pyrite through time. The isotopes are stable and do not change with time, like radioactive isotopes, and the sum of all the sulfur isotopes is therefore always a constant. If a lighter sulfur isotope was sequestered in pyrite, then a heavier sulfur isotope must be left in the seawater sulfate. So the rises and falls in the atomic mass of the sulfur in ancient pyrites over time reflects the amount of pyrite that is being formed.

By the end of the millennium, young Berner had become one of the world's leading scientists, showered with honors and known as the grand old man of geochemistry.[13] Running sophisticated mathematical models of the Earth system and pinning them down with the pyrite data, Berner measured the amount of oxygen and carbon dioxide in the Earth's atmosphere over the past 550 million years. Berner's work showed how the oxygen in the atmosphere had varied between 10% and 30% during this period and enabled understanding of the fundamental drivers of this process. He was also able to show how the carbon dioxide content of the atmosphere had varied in prehuman times, not only providing a global baseline for considerations of the effects of anthropogenic carbon dioxide on climate change but also providing an insight into the processes controlling the composition. And central to these was, of course, pyrite.

Death Zones in the Unfriendly Sea

We had an annual family holiday at the delightful coastal resort of Båstad on the Swedish coast for many years. In the late 1970s we noticed an increase in the number of jellyfish. Apart from being a potential danger to the children in terms of the stings from the tentacles, they washed up on the beach in large numbers and rotted. This made the beach extremely unpleasant. The next year the first of the great blooms of red algae occurred, which made the water poisonous. Carcasses of large cod were scattered on the sands, often with their mouths distended as if they were gasping for air. The algae had rotted and used up all the oxygen in the water, ultimately killing all the fish and creating a dead zone. Unbeknownst to us we had observed the now-classic sequence in the development of marine dead zones. The proliferation of jellyfish was an early signal that the coastal waters were becoming oxygen starved. The other marine animals had moved away or died, leaving the waters clear for the highly adaptable jellyfish. The next year, the oxygen-free waters reached the strand itself, leading to the algal blooms seen along the shore. The basic cause of the oxygen depletion of the coastal waters in this case was the overuse of nitrate fertilizers by Swedish farmers. The nitrate washed into the sea and caused the explosive growth of the red algae. The Kattegatt, the strait that separates Sweden and Denmark, had become one of more than 405 dead zones that now litter the world oceans.

The proliferation of these dead zones has become a major global marine problem. The formation of dead zones is commonly a precursor to the development of the far more dangerous *euxinic* conditions. Euxinic conditions are defined as environments where H_2S occurs in the seawater itself and is not just confined to the sediments. The escape of H_2S into the water column means that more pyrite is formed in sediments, since the normal sedimentary production is complemented by a supply of pyrite formed in the water column itself.

The pyrite buried in sediments fixes the sulfur that sulfur-oxidizing bacteria have not been able to return to the oceans as sulfate. Most pyrite is formed in the upper few centimeters of the sediment where it is exposed to the activities of sulfur- and iron-oxidizing bacteria. We have seen that more than 80% of the sulfide, including pyrite, that is formed in the sediments is reoxidized by sulfur-oxidizing bacteria and returned to the oceans as sulfate.

This is an important reaction. Sulfate itself is not toxic to organisms, but the hydrogen sulfide produced by sulfate-reducing bacteria is extremely poisonous to most forms of life. The sulfur-oxidizing bacteria remove some 240 million tons of poisonous H_2S each year. But for the activities of the sulfur-oxidizing organisms, the whole Earth would smell of rotten eggs.[14]

The formation of pyrite in sediments puts a brake on the potential escape of poisonous H_2S into the oceans. However, the amount of pyrite that can be formed is also dependent on the amount of iron available. In the modern oceans, the iron concentration is extremely low. The reported concentration of dissolved iron has decreased steadily during my lifetime as analytical techniques improved. It was not until the end of the 20th century that the dissolved iron content was established at about 1 nanogram per liter; that is, 1 liter of seawater contains around 0.000 000 001 grams of dissolved iron. It was quickly established that iron is a limiting nutrient for life in the oceans and the bioavailability of iron controls primary productivity. These concentrations are very low, and the modern oceans contain only enough iron to dissolve a tiny part of the H_2S produced in the sediments. The bacteria produce H_2S from the dissolved sulfate in seawater, and there is more than 1 million times more sulfate in the oceans than iron. Much of the pyrite formed in modern sediments is through the reaction of the bacterial sulfide with the iron minerals, which are mainly supplied to the oceans via rivers.

The word *euxinic* derives from the Latin name for the Black Sea, *Pontus Euxinus* or the Welcoming Sea. In fact, this may be a corruption of an earlier name *Pontus Axeinos* as alluded to by Ovid in *Tristia*:[15]

> *Here on the freezing Euxine's shores I stay;*
> *Axine his name, the wiser ancients say.*

Ovid was thoroughly miserable at the time, having been exiled to the Black Sea in 8 CE by the Emperor Augustus for "a poem and a mistake," to quote his own words. He was trying to get back to Rome and wrote a series of poems detailing how awful the Black Sea coast was. The name of this set of poems, *Tristia*, reflects the theme of sadness. *Pontus Axeinos* has been widely translated as the *Unfriendly Sea*. In fact, *Pontus Axeinos* is probably more complex than that and possibly untranslatable since the root word *xeinos* encompasses a concept that is unfamiliar to us: stranger, host, and friend. So *axeinos* is the opposite of this. In any case, modern scholarship would suggest that searching for what the ancients would have understood by a geographic term such as *Pontos Euxeinos* is more difficult than many of the pre-mid-20th-century etymologists would have admitted. For example, it might have been less heroic for Jason and his Argonauts to set sail across a friendly sea in Pindar's famous poem, and Ovid's allusion reflects his miserable banishment to the Black Sea. And, of course, the *axeinos-euxinos* thing is a nice poetic wordplay.

The oxygen concentration of the oceans is delicately balanced. It has long been known that the concentration of dissolved oxygen in the oceans varies with depth, and the oceans are characterized by a midwater oxygen minimum layer that occurs at depths of between 200 m and 1,000 m. Above this layer, the surface water is saturated with oxygen in the atmosphere; below this layer is deep, dense, cold oxygenated water brought in from the ice melting in polar latitudes. Even so, the extent of layers in the oceans that are essentially oxygen-free was not appreciated until the early years of the current millennium. These oxygen-free zones in the oceans are sometimes referred to as *suboxic*, and they are important because they often occur between oxygenated waters and euxinic waters or sulfide-rich sediments.

The minimum oxygen concentration for many higher animals in water is around 1 mg per liter. This compares with the normal dissolved oxygen level in seawater of 3 mg per liter. In suboxic zones the dissolved oxygen levels are less than 1 mg per liter, suffocating many organisms and resulting in the so-called ocean dead zones where higher animals cannot live. The 405 oceanic dead zones occur worldwide and include Chesapeake Bay off the east coast of the United States, the Gulf of Mexico off Louisiana, the northern Adriatic, and coastal waters off South America, China, Japan, and New Zealand.

The increased formation of these dead zones reflects widespread decreases in the dissolved oxygen contents in the present global ocean. These result from global warming.[16] The warming of the oceans results in a deadly combination of increased organic production, decreased O_2 solubility, and intensification of ocean stratification. Warming causes the planktonic organisms that live near the surface of the oceans to increase, which leads directly to an increase in organic matter in seawater and indirectly to an increase in the number of larger organisms. Organic matter in the upper, oxygenated layers of the ocean is oxidized by aerobic microorganisms that utilize oxygen. The oxygen in the water is therefore used up. The problem is exacerbated by the reduced solubility of dissolved oxygen in seawater with increasing temperature. For example, the amount of oxygen dissolved in seawater at 20°C may be 25% less than that at 5°C. Finally, global warming results in increased stratification of the ocean mainly because of the increased temperature contrast between the surface and deep layers. Mixing between layers of ocean waters with different density is limited to physical disturbances such as turbulence caused by storms. Increased density contrasts, especially between the surface layers of the ocean, mean that oxygen cannot reach the deeper waters. The deeper layers become progressively more isolated and oxygen-starved and less habitable to higher life forms such as fish.

The removal of dissolved oxygen from the water column permits H_2S concentrations to build up in the oceans. In the presence of dissolved oxygen, H_2S has a short half-life as microorganisms catalyze the oxidation of this toxic gas to harmless sulfate. At present, wide-scale oxygen-free conditions occur in the oceans where oxygen deficiency is enhanced by upwelling nutrient-rich waters on the eastern oceanic margins between 10° and 40° latitude, such as off the west coast of Peru, Namibia, the north-west margin of the Indian Ocean, the California borderland basins, and the Gulf of California. The nutrients feed large plankton populations, and the organic matter they produce is oxidized by aerobic bacteria, thereby depleting the oxygen concentration in the water. These conditions give rise to dramatic sediment surface discontinuities in sulfide and oxygen concentrations in the sediment and overlying water. The sulfide is some-times kept in check at the sediment–water interface by regional mats of sulfide-oxidizing microorganisms. Local, usually storm-induced, sedi-ment overturns of these sulfide-rich sediments can lead to disastrous results, as described for Walvis Bay in Chapter 6.

Natural oceanic H_2S eruptions are not short-lived, local phenomena but more frequent, regional in extent (up to 20,000 km^2), and longer-last-ing (Figure 8.1).[17] They are triggered by increased intensity of wind-driven coastal upwelling and the passage of a low-pressure weather cell causing

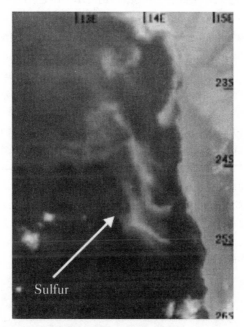

FIGURE 8.1. NOAA Sea-Viewing Wide Field of View images of the ocean off Namibia (20°–23°S) showing turquoise discoloration (appearing white in this image) due to sulfur globules produced as the surface expression of an H_2S erup-tion in March 2001. Modified from Weeks et al. 2004. *Deep Sea Research: Part I. Oceanographic Research Papers*, 51:153–172.

lowering of hydrostatic pressure at depth (through sudden warming of the sea surface and increased rainfall on the hinterland). The result is that euxinia may be more widespread in the present-day oceans than previously thought. It further adds credence to the idea that one side effect of global warming will be increased oceanic euxinia and provides a further alternative explanation for ancient oceanic euxinic events.

The midwater oceanic oxygen minimum layers become especially important where they intersect the continental slope. The combination of suboxic overlying water and sulfide-producing sediments could result, in itself, in the development of extensive areas of oceanic euxinia. This does not seem to be happening at present, although it may have occurred in the past and, if the current global warming trends continue, may occur in the near future.

Pyrite Formation and the Evolution of Life

During global warning events, the oceans are exposed to the deadly trio of increased organic matter production, decreased oxygen solubility, and increased stratification. The result is an increase in global oceanic anoxia and euxinia. These warm intervals have occurred many times in the Earth's long history, and they are commonly associated with the development of extensive regions of H_2S-bearing seawater on the continental shelves. During euxinic events the amount of pyrite in the sediments increases partly as a result of the reduction in the abundance of the aerobic sulfur-oxidizing bacteria, which keep it mainly in check. The natural increase in the amount of pyrite formed in the sediment is augmented by pyrite being formed in the water column itself. However, as mentioned earlier, the amount of iron available to form pyrite in the modern oceans is restricted, and this constrains the possible maximum pyrite concentration. It is really the proportion of iron in the sediment that is in the form of pyrite, rather than as other iron minerals, that increases during euxinic events.

The development of global anoxia has been related to mass extinction events in the geologic past. The greatest mass extinction event in the history of macroscopic life on Earth occurred 250 million years ago: 96% of all marine species were wiped out, as well as 70% of terrestrial species. This great extinction coincided with a major global ocean anoxic event where wide swathes of the continental shelf seas were anoxic and euxinic.[18] The reduction in oxygen availability and the escape of large quantities of poisonous H_2S into the oceans caused this extreme decrease in biodiversity. Terrestrial species, including plants and insects, were affected by increased ultraviolet radiation caused by increased sulfur species in the atmosphere destroying the ozone layer.[19] The whole process was driven by global warming, with sea temperatures reaching 40°C.[20]

During the past 550 million years, for which we have an extensive fossil record, there have been five major mass extinction events when more than 50% of the animal species died out.[21] These occurred at around 450 million, 375 million, 250 million, 200 million, and 66 million years ago. It is often forgotten that the geologic timescale for the past 550 million years is based on the fossil record of the extinction of certain fauna and the introduction of new organisms. Thus the beginning and end of every geologic era, period, epoch, and age is defined by a greater or lesser extinction event. These range from the great extinctions at the end of the three geologic eras to the 100 lesser extinction events marking the boundaries of the 100 geologic ages. In between are the events marking the ends of the twelve more familiar geologic periods, such as the Cambrian and Silurian. There is little evidence that every one of these events was marked by the onset of anoxia and a hike in the sediment pyrite content. Conversely, every increase in the pyrite content of sediments is not related to an extinction event. Even so, there is a correlation between the development of global anoxia and the major mass extinction events. Whether correlation is related to causation in this instance is a moot point. However, arguments based around the potential biological consequences of anoxia, as in the case of the 250-million-year event, can be persuasive.

Each extinction event provides the possibility for other species to fill the vacated ecological niches. Thus mass extinctions may be related to increased rates of evolution. But the relationship with evolution is not clear. It is possible that the new population that replaces the older extinct grouping simply evolves at a different rate. Even so, extinctions are expected to be important factors in driving evolution. In this context, the enhanced pyrite content of sediments, which marks certain major extinction events, is also a marker of a change in the evolutionary history of the Earth.

Pyrite Sulfur Isotope Signatures and the Silver Bullet

Most of Earth history predates the development of extensive animal fossils that have characterized the past 550 million years. During this period it is widely supposed that the oxygen content of the Earth's atmosphere was less than present levels. Indeed, for much of the first 2,500 million years of Earth history, it is thought that there was no significant free oxygen in the atmosphere at all. Most of the early speculation about the nature of the Earth's early atmosphere was based on the English geneticist J.B.S. Haldane's original idea that life must have originated in an anoxic atmosphere, as discussed in Chapter 9. There was little evidence in the rocks. I must say that if you walk over most ancient sedimentary rocks, they do not look much different than sedimentary rocks formed in the oxygenated atmospheres of the past 550 million years. Shales, sandstones, and limestones occur that, where well preserved, look the same as later sediments.

One difference you will notice walking over very ancient terrains is the often extraordinary abundance of iron formations. Iron formations are found in more recent sedimentary sequences but they often make up the main rock type in very old terrains. These abundant iron-rich sediments were mainly produced by chemical sedimentation (Figure 8.2). Currently, these ancient iron-rich sediments provide much of the iron that is powering the Chinese (and Australian) economy. Mountains of these iron-rich rocks are leveled by miners each year and the ore is sent by rail to Port

FIGURE 8.2. Looking into deep time: alternating beds of sedimentary iron oxide and silica in the 2.7-billion-year-old Hamersley range in Western Australia.

Hedland on the north coast, from where it is shipped to China. The 2-km-long iron ore trains transport over 110 million tons of ore to Port Hedland every year, effectively removing annually the equivalent of a ridge of 1,000-m-high mountains 10 km long and 2 km wide. Even so, this is just a drop in the iron ore ocean. The iron ore reserves of Western Australia are around 30 billion tons, which means that miners could continue digging at the same rate for another 300 years. And this is just a fraction of the total of these ancient iron-rich rocks worldwide. Logically the relative abundance of these iron-rich rocks would suggest that the ancient oceans were also iron-rich: that is, iron was not a limiting nutrient on the ancient Earth as it is today. At present, the iron in the oceans is oxidized and precipitates out as iron oxides. In order to get large amounts of dissolved iron in ocean water, the amount of oxygen in the atmosphere needed to be low, thereby preserving the iron in its more soluble, ferrous form.

The main evidence for an ancient oxygen-free atmosphere originally concerned pyrite. Pyrite is an abundant constituent of the sand and pebble sediments that formed in the great Witwatersrand basin in South Africa over 2,700 million years ago. The pyrite was formed in the same way as quartz sand today: by simple mechanical fracturing and erosion of rocks. Today, in our oxygenated atmosphere, the pyrite that is exposed by mechanical weathering processes is oxidized to sulfate and red iron oxides. In the Witwatersrand basin 2,700 million years ago, the fragments of pyrite were deposited in the sediments and were rounded by current action like beach pebbles today. This suggested that the Earth's atmosphere was not oxygenated at that time. The problem was that the skeptics of the anoxic Earth theory could posit environments and processes that might give rise to this apparently detrital pyrite. There was no silver bullet—no incontrovertible piece of evidence that would prove that the atmosphere of early Earth was oxygen-free. The evidence for an anoxic atmosphere in the early Earth, if it existed, was at best cryptic.

As discussed in Chapter 6, bulk sulfur has an atomic mass that is not a whole number because its mass is simply the average of a mixture of sulfur atoms or isotopes that have whole number masses between 32 and 36. The most abundant, at 95% of the total sulfur, is ^{32}S, which has a mass of 32, followed by ^{34}S, which accounts for just over 4% of the total. The average mass of sulfur varies depending on how much of each isotope the sulfur sample contains. The ratios of the two most abundant isotopes in pyrite can be measured, and these show whether the sulfur has been formed by microbiological or volcanic processes. These sulfur isotopes are stable and do not decay radioactively; the sulfur isotope ratio in pyrite is thus preserved over billions of years. The only way of changing it is by reconstituting the pyrite itself through, for example, partial recrystallization under very high temperatures and pressures deep within the Earth.

Although ^{32}S and ^{34}S have been measured in pyrites since the early 1950s, the other stable sulfur isotopes ^{33}S and ^{36}S have not been routinely measured, because ^{32}S and ^{34}S made up over 99% of the sulfur. The technology of the time did not permit analysis of the less abundant sulfur isotopes to a degree of precision as to make them useful. For example, the concentration of ^{36}S is just 0.1% of the total sulfur, and thus analytical precision on the order of 0.001% is required to usefully discriminate between different samples. In addition, the relative concentrations of ^{33}S and ^{36}S are theoretically tied closely to those of the more abundant ^{32}S and ^{34}S, and such measurements made of these less abundant isotopes confirmed this to within about 0.4% precision. That is, the abundance of ^{33}S and ^{36}S in all the pyrites on Earth is a constant fraction of the abundances of ^{32}S and ^{34}S. So it was not worth measuring them. Their concentrations could simply be calculated.

In the 1980s technological advances in mass spectrometry made the precise measurement of these less abundant sulfur isotopes possible. In 2000, James Farquhar, a young isotope geochemist at the University of Maryland, made the amazing discovery that the abundances of the rare ^{33}S and ^{36}S isotopes in sedimentary pyrites deviated from the constant values in rocks older than around 2,500 million years.[22] In rocks younger than this, the sulfur isotope ratios fitted the theoretical relationship. Obviously there was something fundamentally different about the early Earth. Farquhar and his colleagues showed that this effect would occur if there were no significant oxygen in the Earth's atmosphere. They showed that it would occur if there was no ozone layer to protect the Earth from ultraviolet light. But the ozone layer is only there because of the free oxygen in the atmosphere. It was shown that the process causing deviations from the ideal isotope fractionation of the rare sulfur isotopes would only occur if the amount of oxygen in the atmosphere were more than 1 million times less than today's value.

Farquhar discovered the silver bullet proving that the early Earth's atmosphere was oxygen-free—and he did it by analyzing pyrite. His discovery led to a hunt to pin down exactly the moment in Earth history when free oxygen started to build up in the atmosphere. The first results showed that this was between 2,320 and 2,450 million years ago, but subsequent studies have suggested that it might not have been a single event and shorter periods of free oxygen in the atmosphere may have occurred earlier.[23]

Dick Holland, the great Harvard geochemist, called this the Great Oxidation Event.[24] It changed the nature of the Earth forever since it affected not only the surface environment and biology but also the rock cycle and therefore even the deeper processes within the Earth itself.

Still, there were many millions of years between the Great Oxidation Event at 2,300 Ma and the beginning of the evolution of higher animals, which occurred 550 million years ago in what has been called the Cambrian Explosion. It has long been thought that the Cambrian Explosion was made possible by the level of free oxygen in the atmosphere reaching about 10% of the present level. The increased availability of free oxygen was required to power the larger and more complex creatures. It is obvious that the release of free oxygen into the atmosphere must have meant not only that oxygen was being produced in large quantities by photosynthetic primary producers but also that the rate of oxygen production exceeded the rate of removal by reaction with hydrogen sulfide and pyrite. Before that time, the amount of sulfate supplied to the oceans by rivers from the oxidation of pyrite in terrestrial rocks was necessarily small since atmospheric oxygen, which is the main source of pyrite oxidation today, was limited. The lower concentration of seawater sulfate constrained the amount of sulfide that could be produced by sulfate-reducing bacteria and the amount of pyrite formed in the sediments. Counterintuitively, therefore, the increase in oxygen in the atmosphere also increased the potential amount of pyrite produced in sediments and, probably, the frequency and extent of oceanic euxinic events.

Although there was some free oxygen in the atmosphere during the 1,800 million years between the Great Oxidation Event and the Cambrian Explosion, it was far less than there is today. In 1998 Don Canfield of the University of Southern Denmark resurrected an older idea that the oceans during this period must have been sulfidic.[25] Canfield's reasoning was based on the premise that the atmospheric oxygen level was high enough to oxidize pyrite and thus supply some sulfate to the oceans. At the same time, the anoxic conditions that would have prevailed in many of the oceans would have provided an ideal environment for the growth of sulfate-reducing bacteria. Therefore the oceans were probably widely sulfidic, and geochemical evidence was found to support this in analyses of pyrites in sediments of this age in Australia and South Africa. We had a problem with this idea when it was originally suggested in the 1950s, since we did not see masses of pyrite-enriched sediments in ancient geologic terrains.

Canfield concluded in 2012 that he had originally overestimated the extent of sulfidic marine waters after the Great Oxidation Event.[26] He suggested that the nature of the Earth's oceans between the Great Oxidation Event and the Cambrian Explosion can be divided into three periods. Basically the oceans during this time were dominantly iron-rich with infrequent euxinic interludes. However, during the period between 1,800 million and 1,300 million years ago it was more frequently euxinic.

The problem is, of course, the vast swathes of time involved. During the past 550 million years, we know that there have been major fluctuations in the state of the oceans, with periods of widespread euxinia being interspersed with periods of oxygenation. So it is probable that the 1,800-million-year period after the Great Oxidation Event saw a number of vastly different Earth surface environments. The matter is complicated by the limited samples. Many of the pyrite-rich sedimentary rocks preserved during this period were from inshore environments where, even today, one would expect euxinia to occur. Because of plate tectonics, it is rare to find any oceanic sediments older than 200 million years, because older ocean crust is relatively rare and not found in situ. The idea of dominant trends with variable interludes as proposed by Canfield in 2012 seem more in tune with the way the Earth works than hundreds of millions of years of environmental stasis.

Other Earths

I remember as a child in the 1950s listening intently to the most exciting radio program when a BBC reporter was catapulted in a time machine back into the geologic past on Earth. Here he would describe the prevailing environment millions of years ago and fight off marauding trilobites in the Cambrian period or giant jawed fish in the Devonian. When he visited the Precambrian there was nothing. No grass on the land and no fish in the sea. The overwhelming sensation was one of complete silence—a world divorced from any animal life, the only sound being the lapping of the tide and the whistling of the wind in the birdless skies across the treeless rocks.

The Precambrian refers to most of Earth time, from the formation of the planet 4,550 million years ago to the first appearance of abundant shelled animals in the Cambrian Explosion 550 million years ago. The Precambrian thus covers 4,000 million years or four-fifths of Earth history.[27] We can have no real concept of what these enormous periods of time actually represent. As remarked by the English geologist John Playfair in 1802 when introduced to the evidence for the vastness of geologic time by James Hutton, the "father of geology": "The mind seemed to grow giddy by looking so far into the abyss of time."

The 550 million years since the beginning of the Cambrian is called the Phanerozoic, from the Greek for *visible life,* since abundant fossils occur that are large enough to spot with the naked eye—fossils we can see. During these 550 million years we have seen the extinction of the dinosaurs, the origin of land plants, the first animals on land and even, in the last couple of relative seconds the origin of *Homo sapiens.* At the same time there have been great disasters such as the extinction of most species 250 million years ago, asteroid impacts, great ice ages, vast volcanic eruptions, and the continents, riding their tectonic plates, crashing into each other and reforming time and time again. The consequence of this is that if all this could happen in 550 million years, what great changes were wrought to the planet in the eight-times-longer period of time stretching into the depths of the Precambrian? We can look at these immense stretches of time in terms of a series of other Earths, each one very different to the Phanerozoic Earth we inhabit today.

The problem is that as we go back in time, the rock record becomes sparser and more widely scattered and only limited fragments of the ancient planet are preserved. We get mere glimpses of these other Earths. One of the features of scientific advances in the past twenty years has been an increased understanding of the planet's environment during Precambrian time. It has also led to a degree of hubris among researchers in positing static environmental conditions for periods of time well exceeding that of the Phanerozoic eon on the basis of very limited and fragmentary evidence. These interpreted environments are often similar to those occurring in parts of the planet today.

We are like explorers standing on the edge of a vast chasm. We peer into it hoping to see what its depths reveal. The daylight brightly illuminates the first few meters of Phanerozoic time, but then it gets progressively dimmer until only the blackness prevails. There are a few bright spots in the depths of the Precambrian that reveal some data, but they get progressively more difficult to see.

The situation reminds me of the *Voyager* mission in the latter half of the 20th century, which was arguably the greatest experiment of that century. The *Voyager* spacecraft first visited the planets and moons of the solar system and discovered not only extraordinary planetary systems but planetary environments we could not imagine. The researcher into the Precambrian Earth is like my BBC radio reporter traveling back in time to experience and record the nature of these strange environments. And, like the *Voyager* mission, it is likely that he will discover other Earths that we cannot imagine at present.

One of the key materials in the rock record is pyrite, which lights up the past in the same way as the myriads of its nucleation bursts illuminate recent muds, casting persistent microscopic sparks into the eternal darkness of past eons. Pyrite is relatively resistant to the ravages of geologic time. It withstands temperatures up to several hundred degrees centigrade, where it tends to lose its sulfur, and reforms when the temperature decreases. It does not change under extremely high pressures. It tends to fracture rather than bend and deforms through recrystallization. Even so, it retains large quantities of the original material within these new crystals as cores or fossils of the original sedimentary pyrite. The lack of physical equilibration of pyrite is also reflected in a resistance to chemical and isotopic changes. Ancient pyrite crystals are thus like little bottles of ancient sedimentary environments cast into the sea of deep time, carrying messages to us about what it was like then. They can tell us about the chemistry of the oceans and atmosphere, the nature of the contemporary life forms, and how the Earth system worked and evolved. As such, pyrite is the primary target of research into the ancient Earth, and much that we know, or think we know, about the evolution of the Earth and the nature of ancient environments comes from studies of ancient pyrites. A pyrite crystal is a miniature window into ancient worlds.

The oldest sedimentary rocks preserved on Earth are the 3,790-million-year-old clastic sedimentary rocks of western Greenland. These rocks contain pyrite with a sulfur isotopic signature suggesting both a sedimentary and a microbial source. At one time these constituted the oldest rocks on Earth, which caused some consternation. The problem was that clastic sedimentary rocks are the weathered fragments of other rocks that have been eroded and deposited in the oceans or along the coast like on sandy beaches today. These clastic sediments must ultimately be sourced in even older igneous rocks that were created at or near the original planetary surface. Subsequently, older igneous rocks have been found in Greenland with ages of 3,820 million years, and the oldest rocks of all are probably the 4,030-million-year-old Acasta gneisses, highly deformed and altered rocks of mainly igneous provenance, in the Northwest Territories of Canada.

The Greenland sedimentary rocks have been altered by heat and pressure and are severely deformed, so a detailed reconstruction of the sedimentary environment is difficult. By contrast the sedimentary rocks in the 3,500-million-year-old rocks of the Pilbara region in northwest Australia are remarkably well preserved. They were formed in shallow lagoonal environments in a partly enclosed bay fed by volcanic sulfate. They include the oldest fossils known on Earth, single-celled microorganisms preserved in silica. They also include sedimentary pyrite with isotope signatures showing a microbiological origin for the sulfur in the pyrite.

It seems that one of these early Earths was a strange environment with a small Sun blasting the surface with intense ultraviolet light. The atmosphere was not breathable to us since it contained no free oxygen and was mainly carbon dioxide and nitrogen. The land was bare of any life and was just rocky and barren with no grass, trees, or soil. The lakes and rivers were free of sulfate and were probably rich in dissolved carbon dioxide, rather like soda water without the fizz. The waters were probably mostly pristine and pure as they were free of macroscopic life. The oceans appear to have contained large concentrations of dissolved iron for extended periods, and thus they were probably acidic and a translucent, inorganic pale green color unlike the deeper biological green of the life-enriched seawater of today.

Overall we currently know little about what the oceans were like, although various theories have been proposed, often based on analyses of contemporary sedimentary pyrite. Unfortunately the pyrite was formed within the sediment and was thus not in direct contact with ocean water or the atmosphere. Although pyrite constitutes a miniature window into these distant Earths, it refracts the information, rather like seeing a distorted image of the world when looking through an emerald. Our science, although it has progressed immeasurably during the past century, is not yet sufficiently advanced to enable us to unravel the information about the early Earth provided by pyrite. However, we are beginning to see glimpses of the meaning of the traces of the myriad elements contained in ancient pyrites, starting to understand the implications of the variations in the isotopic compositions of iron and other nonconventional isotopic sources in pyrite, and beginning to unravel the meaning of the traces of organic derivatives associated with pyrite.

Notes

1. B.B. Jørgensen. 1982. Mineralization of organic-matter in the sea bed—the role of sulfate reduction. *Nature*, 296:643–645.

2. S.H. Bottrell and R.J. Newton. 2004. Reconstruction of changes in global sulfur cycling from marine sulfate isotopes. *Earth Science Reviews*, 75:59–83.

3. R.J. Andres and A.D. Kasgnoc. 1998. A time-averaged inventory of subaerial volcanic sulfur emissions. *Journal of Geophysical Research-Atmospheres*, 103:25251–25261.

4. H. Elderfield and A. Schultz. 1996. Mid-ocean ridge hydrothermal fluxes and the chemical composition of the ocean. *Annual Review of Earth and Planetary Sciences*, 24:191–224.

5. K. Sugawara, T. Koyama, and A. Kozawa. 1953. Distribution of various forms of sulphur in lake-, river- and sea muds (I). *Journal of Earth Science of Nagoya University*, 1:17–23; 1954. Distribution of various forms of sulphur in lake-, river- and sea muds (II). *Journal of Earth Science of Nagoya University*, 2:1–4.

6. C.A. Kuhr, H. Von Wolzogen, and N.S. Van Der Vlugti. 1934. The graphitization of cast iron as an electrobiochemical process in anaerobic soils. *Water*, 18:147.

7. See, for example, R.W. Hyman, R.P. St Onge, H. Kim, et al. 2012. Molecular probe technology detects bacteria without culture. *BMC Microbiology*, 12. http://www.biomedcentral.com/1471-2180/12/29.

8. R.A. Berner. 1971. Worldwide sulphur pollution of rivers. *Journal of Geophysical Research*, 76:6597–6600. E.K. Berner and R.A. Berner. 1987. *The Global Water Cycle: Geochemistry and Environment* (Englewood Cliffs, NJ: Prentice Hall).

9. For example, a 47% increase in the riverine supply of sulfate would lead to a 1% increase in seawater sulfate concentrations in less than 20,000 years. Even though the total mass of sulfate in the oceans is huge, at around 4×10^{21} g, this is a potentially planet-changing rate of change if maintained over geologic time periods.

10. J.J. Ebelmen. 1845. Sur les produits de la decomposition des espèces minerales de la famille des silicates. *Annales des Mines, 4eme Series*, 7:3–66; 1847. Sur la decomposition des roches. *Annales des Mines, 4eme Series*, 12:627–654. Ebelmen died in 1852, only thirty-seven years old. He had a remarkable career in these short years, especially in the fields of synthesizing gems and porcelain manufacture. His is one of the seventy-two names inscribed on the Eiffel Tower. These papers were ignored by the geochemical community until 1996, when Berner and Maasch rediscovered them. Similar conclusions had been reached, apparently independently, by the great American geochemists Harold Urey and Bob Garrels in the mid-20th century.

11. $2Fe_2O_3 + 8SO_4^{2-} + 16H^+ \underset{2}{\overset{1}{\rightleftharpoons}} 15O_2 + 4FeS_2 + 8H_2O$

    Reaction 1 is the formation of pyrite through sulfate reduction, and reaction 2 is the oxidation of pyrite producing sulfate and acid.

12. $CO_2 + H_2O \underset{2}{\overset{1}{\rightleftharpoons}} CH_2O + O_2$

    Reaction 1 represents photosynthesis where carbon dioxide and water are converted to organic matter, represented as CH_2O and oxygen. Reaction 2 represents the mineralization of organic matter to carbon dioxide. This reaction is obviously carried out by aerobic organisms. Sulfate reducers perform the same process but here the oxygen comes from SO_4 leaving sulfide behind.

13. See R.A. Berner. 2004. *The Phanerozoic Carbon Cycle* (New York: Oxford University Press), 150pp, for a summary of this work.

14. Troposphere mass = 4.1×10^{18} kg. 240×10^6 tons H_2S per year is equivalent to 0.06 ppm H_2S in atmosphere. The human nose detects 0.005 ppm H_2S. Even without the activities of the sulfur-oxidizing bacteria, H_2S reaching the atmosphere is oxidized to sulfate by reaction with atmospheric oxygen. However, the back-of-the-envelope calculation here takes into account only the amount of H_2S reaching the atmosphere in just one year. If this were to continue for any length of time, the oxygen content of the atmosphere would begin to decrease noticeably.

15. Ovid. *Tristia*. IV: 4–56.

16. The current rate of decrease in dissolved O_2 in the oceans is model dependent but ranges between 2 and 4 micrograms of oxygen per liter per year. Since the current average oceanic dissolved O_2 concentration is about 3 mg per liter, it would appear that all the oceanic O_2 might be removed in less than 1,000 years if the present rate were to continue. This is obviously an incorrect figure since it does not take into account other factors, such as ocean mixing times and ventilation. However, it does suggest that the development of oceanic anoxia on a global scale could occur relatively rapidly from a geologic standpoint. This is consistent with information about the development of global oceanic anoxic events in the past. See, for example, R.F. Keeling, A. Kortzinger, and N. Gruber. 2010. Ocean deoxygenation in a warming world. *Annual Review of Marine Science*, 2:199–229. R.J. Matear, A.C. Hirst, and B.I. McNeil. 2000. Changes in dissolved oxygen in the Southern Ocean with climate change. *Geochemistry, Geophysics, Geosystem*, 1(11), 1050, doi:10.1029/2000GC000086. L. Stramma, G.C. Johnson, J. Sprintall, and V. Mohrholz. 2008. Expanding oxygen-minimum zones in the tropical oceans. *Science*, 320:655–658.

17. S.J. Weeks, B. Currie, A. Bakun, and K.R. Peard. 2004. Hydrogen sulphide eruptions in the Atlantic Ocean off southern Africa: Implications of a new view based on SeaWiFS satellite imagery. *Deep Sea Research: Part I. Oceanographic Research Papers*, 51:153–172.

18. P.B. Wignall and R.J. Twitchett. 2002. Extent, duration, and nature of the Permian-Triassic superanoxic event. *Geological Society of America Special Papers*, 356:395–413.

19. L. Kump, A. Pavlov, and M.A. Arthur. 2005. Massive release of hydrogen sulfide to the surface ocean and atmosphere during intervals of oceanic anoxia.
Geology, 33:397–400.

20. Y. Sun, M.M. Joachimski, P.B. Wignall, et al. 2012. Lethally hot temperatures
during the early Triassic greenhouse. *Science*, 338:366–370.

21. D.M. Raup and J.J. Sepkoski Jr. 1982. Mass extinctions in the marine fossil
record. *Science*, 215:1501–1503. The problem is that most life on Earth is microbiological and there is limited evidence that the mass extinction events had
any effect on microorganism speciation. By contrast, it might be expected that
species-specific animal pathogens die out with their hosts and at least the numbers of aerobes decrease as anoxia spread.

22. J. Farquhar, H.M. Bao, and M. Thiemens. 2000. Atmospheric influence of
Earth's earliest sulfur cycle. *Science*, 289:756–758.

23. Euxinic conditions have been suggested at 2,500 million, 2,660 million, and
2,460 million to 2,590 million years in West Australian and South African rocks
(summarized in R. Raiswell and D.E. Canfield. 2012. The iron biochemical cycle
past and present. *Geochemical Perspectives*, 1(1), 232pp). Euxinia require sulfate,
which might be related to elevated atmospheric oxygen levels and independent
evidence for free atmospheric oxygen, has been found for the 2,500-million-
year event. This is some 170 million years before the Great Oxidation Event. The
problem with relating euxinia to atmospheric oxygen is that elevated seawater
sulfate can be locally supplied by volcanic sulfur dioxide and not only by pyrite
oxidation.

24. Strictly this term was coined by R. Rye and H.D. Holland. 1998. Paleosols and
the evolution of atmospheric oxygen: A critical review. *American Journal of
Science*, 298:621–672—but it will always be associated more with Dick Holland
because he went on to write a series of epoch-making books about the evolution
of the atmosphere and oceans of the Earth.

25. D.E. Canfield. 1998. A new model for Proterozoic ocean chemistry. *Nature*,
396:450–453. Earlier H.L. James had found that the deeper levels of the
1,800-million-year-old oceans were sulfidic (H.L. James. 1951. Iron formation
and associated rocks in the Iron River District, Michigan. *Geological Society of
America Bulletin*, 62:251–266). James and Sims went on to suggest extensive
anoxic oceans at his time (H.L. James and P.K. Sims. 1973. Precambrian iron-
formations of world. *Economic Geology*, 68:913–914), and this was suggested by
Canfield in 2008 for the period between 2000 and 1,000 million years.

26. R. Raiswell and D.E. Canfield. 2012. The iron biochemical cycle past and present.
Geochemical Perspectives, 1(1), 232 pp.

27. On a jingoistic note I cannot resist pointing out here that most of geologic
time is named after Wales. The great Cambrian period that represents the
time between the beginnings of abundant visible fossils at 541 million years
ago lasted 56 million years and is named after *Cambria*, the Roman name for
Wales. This was followed by the Ordovician and the Silurian periods, which

took time along another 66 million years up to 419 million years ago. These periods were named after Celtic tribes, the *Ordovices* and *Silures*, which lived in Wales. The legendary King Arthur was a king of the *Silures*. So Welsh names cover over a fifth of Phanerozoic or fossiliferous time. But then there is the Precambrian—also named after Wales—which covers the 4,000-million-year period from the formation of the Earth 4,550 million years ago to the beginning of the Cambrian. This means that over 90% of geologic time is named after Wales.

9

Pyrite and the Origins of Life

IF YOU HAVE been reading this book since the beginning, you will not be surprised by now to find that you have come across a chapter documenting the involvement of pyrite in the origin of life. This is because you will have read in this book how pyrite has been at the root of many fundamental discoveries about the nature of our world. So you do not suffer more than eyebrow-raising surprise and maybe a gentle throat-clearing in learning that pyrite is contributing to our current understanding of the origins of life.

By contrast, if you have dived in at Chapter 9 you probably look at the title of this chapter with disbelief. After all, what could be the connection between a common glitzy mineral and the origin of life? The more diligent reader will have already learned that pyrite formation is intimately associated with biology because most of it is produced by bacteria that extract their oxygen from sulfate and produce hydrogen sulfide. This relationship is so overweening today that pyrite formation controls many fundamental aspects of the Earth's environment. So what happens if we extend this line of inquiry back to the beginnings of geologic time? We have already seen that the characteristics of ancient pyrite are one of the main sources of information about the nature of the early Earth. The consequence of this is that we know quite a bit about the relationship between pyrite and early life on Earth. In this chapter, we further explore this and review the laboratory work that implicates pyrite itself in the original syntheses of the self-replicating biomolecules that assembled to produce Earth's first life forms.

Abiogenesis

The thesis that life developed from nonbiological chemistry is a very old idea stretching back through Anaximander in 6th-century BCE Greece to the Vedic writings of ancient India around 1500 BCE[1] and is often called *abiogenesis*. It was systematically conceptualized by Aristotle in 350 BCE as *spontaneous generation*, the idea that animals sprang, wholly formed, from mud. This reached its apotheosis in the 1642 recipe for making mice by Johannes Baptista van Helmont: incubate a flask of wheat and old rags in a warm, dark closet and mice will appear. In 1862, the French bacteriologist Louis Pasteur hammered the final nail in the coffin of spontaneous generation in a famous experiment that proved that spontaneous generation did not occur.[2] Spontaneous generation was a warped theory of animals springing already formed from inorganic sources, such as mud. In fact the originators of the idea of abiogenesis did not propose this. Anaximander, for example, appeared to suggest that germs of life were formed in mud and that these evolved into complete organisms. This idea is not much different from the modern concept where Anaximander's germs are organic molecules with the capability of reproducing themselves.

If we were to take a straw poll of scientists today, they would generally include the origin of life as the greatest scientific quest of our era. It has settled in the center of our communal consciousness since the discovery that deoxyribonucleic acid (DNA) is the molecule that enables heredity to be transmitted through generations of organisms. The structure of DNA was resolved by X-ray diffraction analysis, and, as I discussed in Chapter 4, pyrite played an important role in the development of that technique. The classic photograph of Crick and Watson manipulating their man-sized model of the DNA molecule, with all of its sticks and balls and platens, seemed to suggest that the manufacture of DNA is just a matter of engineering. Since DNA is key to modern life, it invited the idea that the creation of life in a test tube was just around the corner.

The problem is that the Crick and Watson model is just that: a model of the reality of DNA. The molecule is actually extremely small on a human scale, with myriads of smaller molecules coiled within coils and highly complicated. Although the structural analysis of DNA revealed that the genetic code was contained in just four relatively simple organic molecules called bases, the number of possible combinations of these four bases increases by a power law related to the number of these molecules. So, for example, four bases have 256 different combinations whereas five bases have 625. In one typical DNA molecule there might be 200 million of these bases. The human genome, for example, is spread over twenty-three such DNA molecules or chromosomes, so there is a virtually infinite number of possible combinations. Some of these combinations work, in the sense that they code for a viable life form, but most do not. In fact life today is even much more complicated than that. The genetic code needs to be transmitted to the proteins that do the work, and the proteins need to be synthesized as well.

More significantly, it is often overlooked that the DNA we see in modern organisms is the result of 4 billion years of evolution. We know it has changed in that time since the fossil record shows a timeline of evolution from single-celled organisms to multicellular animals and plants. The DNA of a dinosaur must have been different from the DNA of an oyster. Wildly different organisms do share a vast number of genes, however. Sir Robert May in his inaugural address to the distinguished Fellows as President of the Royal Society in 2000 said, "It has come to my attention that we share half of our genes with those of a banana. And looking around the room today, I am not a bit surprised."

The bits of DNA that we do not share with the banana distinguish us as humans. The DNA we see today, which is the only DNA we know of, is not the same as the DNA of the earliest organisms. The characteristics of organisms are defined by their DNA: since these characteristics have changed over time, so DNA must have also changed. There are two powerful feedbacks in modern evolutionary theory. First, organisms react to the continuously changing environment so that only the fittest survive; the DNA of those organisms is then selected for preservation and further evolution. Second, occasional errors in copying DNA and the splicing in of bits of DNA from other organisms during an infection, for example, lead to the development of further characteristics in organisms that may or may not be more successful in the local environment. Sulfate-reducing bacteria, which are key to producing pyrite on Earth, can utilize the greatest number of alternate metabolites in the microbial world. In other words, their DNA contains numerous gene sets that enable them to carry out processes other than sulfate reduction if sulfate is not available. These gene sets have been introduced into their DNA in the past; they have not done the sulfate-reducers any harm, and they have helped them survive the lean times when sulfate was scarce.

The consequence is that DNA is like a tape in a tape recorder, recording the evolutionary history of the organism and showing its provenance and relations. In fact, it appears that the DNA of modern organisms contains bits and pieces from their whole evolutionary history. Mathematical analysis of the differences in the ordering of the bases in modern DNA, then, enables the evolutionary relationships of modern organisms to be deciphered. This has led to a number of less-than-savory commercial enterprises offering to code your DNA and tell you where your ancestors came from. Drawing these mathematical relationships between all the modern DNAs produces a tree-like form with progressively fewer species as you travel toward the root, analogous to fewer branches in a bush. Finally, there is just one, the root of the bush—the common ancestor of all modern life. The common ancestor does not exist today, and it is really just a mathematical construct based on extending known DNAs back in time.

DNA is a polymer, like nylon, plastics, and rubber. It is a very large molecule in terms of its atomic weight, but it is not the largest known. The largest molecules are mostly human constructs: there is a cogent argument that the largest single molecule on Earth is a dump-truck tire. Polymers are molecules that have no chemical limit to their size: they consist of simply repeated additions of the basic constituent molecules. So the formation of modern DNA, which may contain hundreds and millions of molecules, has resulted from incremental additions and alterations over the eons of geologic time. The same process was originally invoked to explain the evolution of whole organisms or complex parts of organisms like the eye. Of course we know now that small changes over long periods of time produce large variations.

I am reminded of an American tourist who visited the fête held every summer by the dean in the garden of the Old Rectory in our village in west Wales. He was amazed by the lawn, which was perfectly green, with fine soft grass and a springy feel underfoot. He asked the dean the secret of making a lawn of this quality: "Well, the key thing is regular mowing. You should mow the grass once a week. If you do that every week for 300 years you'll get a lawn like this!"

The only DNA we know is modern DNA, and this DNA is the result of 4 billion years of evolution, or mowing, as the dean would suggest. The mowing in this case is the removal of those DNA combinations that have produced less successful organisms: that is, Darwinian evolution. Making DNA is not the target of origin-of-life research. The scientific community is not looking into ways of achieving spontaneous creation, of adding some mysterious potions to a test tube and having fully fledged mice jumping out.

We can see that research into the origin of life cannot meaningfully start with modern DNA. It seems extremely unlikely that DNA had to be formed first in order for life to be created. The original molecules of life were probably far simpler than modern DNA. They did have two of the fundamental attributes of DNA, however: they were carbon-based and they had the possibility to replicate themselves. DNA is a biopolymer, and the manufacture of polymers involves adding molecules together in a specific order and orientation. It is often carried out with the aid of catalysts, substances that help the process but can be recovered after it is complete. In the case of the origin of the complex biological molecules such as the precursors to modern DNA, pyrite has been proposed as a catalyst.

The search for the origin of life is further complicated by the fact that although all life today is based on DNA, this does not mean that DNA was the first molecule to reproduce itself. All it means is that all life today is derived from an ancient common ancestor that was DNA based. This common ancestor to modern life may have been just one life form on the ancient Earth: it just happened that the DNA molecule (or its precursor) was more successful than its competitors. In other words, life may have originated with a variety of basic biochemistries, and it may have started many times in different places. Hence the title of the chapter: the "origins" of life rather than the "origin" of life.

Carbon appears to be fundamental to life on Earth because the carbon atom can join with other carbons at relatively low temperatures to form complex chains, rings, and three-dimensional molecules necessary for a life form. If any other readily polymerizing element was involved during the origins of life, such as silicon for example, it has been outcompeted by carbon-based life forms and has left no trace on Earth that has as yet been recognized as a life form. The first stage in the formation of life is then to create molecules in which carbon atoms are bonded to other carbon atoms, rather than oxygen or hydrogen for example. The formation of carbon–carbon bonds from inorganic carbon species such as carbon dioxide or methane at ambient temperatures is a nontrivial process. It is in basic processes like these where pyrite may have played a key role.

The Iron-Sulfur World

We do not actually know much about the nature of the Earth's earliest environment where life was originally developed. The problem is with the huge amounts of time involved. The concept of the length of geologic time is neatly encapsulated in John McPhee's phrase *deep time*.[3] Some appreciation—as far as our human experience allows—of the vastness of geologic time is needed in this chapter since it has become increasingly popular in recent scientific literature to make sweeping generalizations about the Earth and its deep past based on very few data points.

Various attempts have been made to transmit the concept of deep time, and one of the most common is representing the whole of geologic time in one day. Humans, in our present form, have been around on the planet for about 200,000 years—we appeared in the last 4 seconds of the geologic day, at 4 seconds to midnight. During this time we have experienced great ice ages interspersed with periods of extreme warming. Sea level has varied by several hundred meters, with much of the land we know now underwater at times and land being exposed where there is now sea. We have experienced large meteorite impacts (although not catastrophic ones), volcanic eruptions, and earthquakes. The atmosphere has dramatically changed with carbon dioxide and oxygen fluctuating in concentration. If all this can happen in 200,000 years (or 4 seconds of our geologic day), what changes occurred in 200 million years, about an hour in our geologic day? There were 4 billion years between the formation of the Earth and the first widespread appearance of visible fossils, which occurred at about 20 minutes to 10 in the evening of our geologic day. The problem is that as we go back in time, the precision of our present methods of dating rocks decreases. In particular, in the period before 1 billion years ago, an uncertainty of 2 million years means a precision of greater than one-hundredth of 1 percent in measurement. With recent technical advances precisions approaching this level can be reached. However, this period of time still represents ten times the length of time we humans have been on the planet. Thus the environment in which life started on Earth cannot be known with any certainty. We can make generalizations about the conditions on the planet during its early years, but we know nothing about any specific period of time when the first life may have originated.

To place this in context, I think that the chemical reactions that produced the first sustainable organic molecule on Earth could not have taken much time. This is because these chemical reactions required a dynamic environment to bring the reactants together in sufficient concentrations at the right time. And, by definition, such a system is unstable and does not last long. The idea of these prebiotic organic molecules being formed in an instant seems unavoidable, and tracking down this instant in the vast wastes of geologic time appears impossible at present. Once the first

molecules were formed, they would be able to reteact to form more complex and more efficient architectures. In fact, it is probably impossible
to get rid of them. The history of the Earth shows that once a planet is
infected with life, it will never recover. It is stuck with life as long as the
planet exists. Fate may throw huge asteroids at it, gigantic volcanic eruptions may cover its surface, and earthquakes may occur that are so large
that the planet rings like a bell, but life finds a way of persisting somehow.
Occasionally, one species does so well that it dominates the biosphere and
its waste products kill off many, if not most, of the other species on the
planet. Ultimately, this species self-destructs and life continues to spin its
merry way through time.

However, the best we can do to define the environment of the origin of
life on Earth is to define certain general conditions that might have been
prevalent around that time. The problem, then, is that these conditions
might have occurred many times in the past, so if these were prerequi-
sites for the origin of life and the reactions involved happened quickly, life
could have started many times: it would have had many origins.

The Oldest Pyrite

The vagueness of our knowledge about the time any rock sample from the
early Earth represents is compounded by the increased scarcity of rocks
as we go back in time. Indeed, the first period of Earth history, called the
Hadean after the Greek Hell, was originally defined as the period when no
rocks have been preserved. It is generally accepted to represent the first
600 million years of the Earth, or the time until 3 o'clock in the morning
of our geologic day. Rocks have been found that are 4 billion years old in
the Slave Province of northern Canada, and tiny 4.4-billion-year-old zircon
crystals, formed just 70 million years after the Earth was finally constituted, have been preserved in younger Australian rocks.[4] Unfortunately
these crystals, derived from just half an hour after midnight on the morning of our geologic day, were formed deep in the Earth's crust and tell
us little about the environment where life may have originated. Even the
rocks formed 4 billion years ago do not tell us much about the surface
environment of the Earth during the first three hours of our geologic day.

Most of geologic time is not recorded in the rock record, even in
recent times. Sedimentary rocks are laid down in layers of various thickness called beds. These are separated by distinct surfaces called bedding
planes, as shown in Figure 8.2 in Chapter 8. Even in modern sedimentary

rock sequences, the rock itself may represent sediments that have been deposited in a few thousand years and often through some catastrophic event such as a flash flood or a mud slide that is over in less than a day. By contrast, the bedding plane represents an indefinite period of time where no sedimentation occurred or where sediments were eroded. The beds of the 2,700-million-year-old iron sediments in the Hamersley range, shown in Figure 8.2 in Chapter 8, were deposited as chemical sediments at a rate of around 180 meters of sediment every million years.[5] This seems like a lot until we realize that this is just 0.2 mm of sediment each year. In fact much of the iron formations are made up of very thin bands averaging about 1 mm thick. If these bands represent annual sediments, you can see that the rocks record only one-fifth of the actual elapsed time: at least four-fifths are missing. The missing time is in the bedding planes. The consequence is that the rocks do not record most geologic time.

Near where I sit writing are rocks that were formed around 200 million years ago when Wales was in the center of a great desert larger than the Sahara. In this desert there were dry gulches like the Mojave desert of the western United States and, like the Mojave dry gulches, they were susceptible to flash floods. A thunderstorm dumps its rainwater up to 50 km away and the water rushes across the desert, through the canyons on its way to the sea. The first you hear is a distant rumbling as if there's a herd of wild horses coming over the horizon. When you sense this, you must scramble up the dry gulch walls because, within minutes, the sandy gulch is full of a raging torrent of white-flecked, brown water crashing and heaving, carrying away everything in front of it with huge boulders and whole trees thrashing in the spume. Every year, at least when I was working there, someone was killed wandering up a peaceful dry gulch under a beautiful clear blue sky by these flash floods. When the flood subsides it leaves a deposit, quite thick in places, of sand and boulders and trees. These sediments have been formed in an hour: they record what happened during this hour, leaving the remaining 8,759 hours of the year unrecorded. The same thing happened 200 million years ago in Wales, and the sedimentary rocks that are preserved provide a record of just a fraction of the time period during which they were formed.

When we delve into the deep past of the Earth, the samples of rocks become so rare that there may be only one location covering 100 million years of time. And the rocks in that location may record less than 1 million years of Earth history. By around 3.85 billion years, or 20 past 3 in the morning of our geologic day, we come to a time that geologists generally suppose is the beginning of the continual rock record. Note that this rock record is *continual*—forming a sequence in which the same event is recurring—rather than *continuous*—forming an unbroken whole, without interruption. In fact sedimentary rocks from around 3.5 billion years are only known from two locations on Earth, in Western Australia and southern Africa.

The 4-billion-year-old rocks in Canada contain pyrite, but its age is uncertain and it may have been introduced during a later event. These ancient rocks probably did contain pyrite when they were formed since similar, younger rocks always contain pyrite. However, we cannot be sure at present and, more important, we cannot distinguish any pyrite grains that were formed 4 billion years ago from later ones. Sedimentary rocks in west Greenland probably contain the recognized oldest pyrite grains on Earth. Some of these pyrite grains have been directly dated to 3.85 Ga and have provided the basis for a novel method for dating sedimentary rocks based on the isotopic compositions of gases trapped in mineral impurities in the pyrite.[6] The pyrite grains act as tiny time capsules, armoring minute inclusions and preventing the escape of the radiogenic elements used for dating. So we know that pyrite has been forming on the Earth's surface for all of recorded geologic history.

The pyritic Greenland sediments are important because they prove that water was present on Earth's surface at that time. Sediments are formed in water, and these ancient Greenland sediments include conglomerates from rough pebbly beaches and iron formations where iron has been chemically precipitated from water. It is likely that the atmospheric pressure 3.85 billion years ago was not substantially different from today, so the existence of water shows a surface temperature of less than 100°C. Admittedly, this is just a snapshot of the Earth at this point in time, but it seems that water has been present on the Earth ever since. These rocks also appear to show the earliest evidence for life.[7] However, no fossils have been reported from these ancient rocks in Greenland, and the evidence for life is indirect, related to the isotopic composition of carbon in the mineral graphite.

Interpretations of the Greenland rocks are complicated by their history: they have generally been heated to over 500°C and enjoyed pressures in excess of 1.2 GPa or 12,000 atmospheres. At these temperatures and pressures deep in the Earth's crust, even granite softens and begins to flow and deform like pastry dough. This is why any original organic carbon has lost its volatile constituents and is now pure carbon, graphite, and why no fossils of early life from this time have been preserved. The abundant pyrite grains in the rock do not show evidence for biological processes, although they have retained that characteristic isotopic signature, discussed in Chapter 8, that shows that the contemporary atmosphere was oxygen-free at that time.

In summary, therefore, we know that the origin of life on Earth occurred before 3.85 billion years ago. The rock record shows that water was present on the Earth's surface at this time and that therefore the surface temperature was within the range that is occupied by life at present. In other words, organic, carbon-based life was possible, and the carbon isotopes in the ancient graphite are consistent with this. These earliest rocks also contain abundant pyrite, so we know that pyrite was formed in sediments 3,850 million years ago. However, the sulfur in this ancient pyrite does not contain any evidence for a biological source of sulfur. Most probably volcanic submarine hydrothermal vents, as described in Chapter 5, produced the pyrite.

The Frankenstein Experiment

As noted previously, the 3.85-billion-year-old pyrite grains have retained the isotopic signature that demonstrated that the Earth's atmosphere did not contain any free oxygen at that time. A lack of oxygen is widely thought to have been originally proposed as a prerequisite for the origin of life on Earth by the Russian biochemist Alexander Oparin in 1924.[8] Like most teachers, I have taught students for years that Oparin was the fount of this wisdom. I have since looked at Oparin's original paper in detail and was surprised to find no mention of this. Indeed, in 1924 Oparin seemed to assume that free oxygen was around in the atmosphere of the young Earth when abiogenesis occurred and was a prerequisite for some of the original reactions involved.

It appears that, in fact, the idea of life originating necessarily in an Earth where there was no free molecular oxygen was first proposed by the English aristocrat and Marxist geneticist J.B.S. Haldane in 1929.[9] It is likely that both Haldane and Oparin were influenced by an unfinished 1883 work on science by the Marxist philosopher Friedrich Engels,[10] which inspired modern dialectical materialism. The idea is that since oxygen destroys organic matter (think about burning), the presence of oxygen was incompatible with the formation of organic matter. By 1936, some seven years after Haldane had published his theory, Oparin had changed his mind and the conventional view of Oparin's ideas was created. I am afraid our view of the modern history of abiogenesis has been refracted and distorted by 20th-century politics and philosophy and the occasional war and revolution. Science became a projection of state policy and political philosophy. Perhaps it still is; perhaps it always was.

This historical haze was penetrated in 1953 when a young doctoral student, Stanley Miller, studying with Nobel Prize winner Harold Urey at the University of Chicago, put the Haldane theory to the test. They produced an electric spark—to represent lightning—in a vessel containing water and an atmosphere of methane, ammonia, hydrogen, and carbon monoxide but no oxygen, which represented the likely composition of an early Earth atmosphere. The product was an organic tar that contained a set of amino acids, the chemical building blocks of proteins. This experiment is reminiscent of Dr. Frankenstein putting an electrical charge into bits of humans and producing a monster—Mary Shelley was prescient in this regard. The experiment was so successful that later Miller added sulfide to the mix and produced an even more realistic set of amino acids, suggesting that sulfur was an important component of these early processes.[11]

The upshot of the Miller–Urey experiments was twofold. First, it showed that experiments could be carried out on abiotic syntheses under putative early Earth conditions. It is possible that we will never be able to prove exactly how and when life started on Earth, not least because of the lack of any rocks from that time. However, it may well be that we will be able to show how life could have started on Earth through experimentation under reasonable approximations to contemporary conditions. Second, it demonstrated that some of the building blocks of life could be abiotically synthesized on the primitive planet. Since that time Saturn's satellite Titan has been visited by the *Cassini-Huygens* spacecraft and huge seas of methane and other hydrocarbons have been discovered on that moon. This provides some sustenance to the idea that organic chemicals could have been present on the primitive Earth. Indeed, many researchers now consider that the Earth was shrouded in a methane haze for much of its early history.

In summary, therefore, the weight of current scientific opinion is that it is possible that life on Earth originated at a time when there was no oxygen in the atmosphere, there was water on the surface, and the basic building-block organic chemicals were present. All that is needed now is some means of putting these building blocks together in a molecule that can react to reproduce itself.

Pyrite and the Synthesis of Biomolecules

After the Frankenstein experiment of young Stanley Miller, we were left with the building blocks of proteins but no way of turning them into living matter. This is done in organisms today by DNA, which codes for how these amino acids should be assembled into the bits and pieces of biochemistry that make life work. However, the building blocks of DNA are not amino acids but, as the name *deoxyribonucleic acid* suggests, nucleic acids. These nucleic acids are very different and more complex molecules than simple amino acids. They are not produced in the Frankenstein experiment but appear to require a more complex sequence of reactions.

One way of looking at the nucleic acid polymers like DNA is as templates that control how amino acids are linked in a manner that makes them useful as a protein. We are now presented with a chicken-and-egg situation: we need the nucleic acid to form the protein but we first need something to form the nucleic acid, which is chemically more complex. One way around this was first suggested in 1949 by another Marxist philosopher scientist, J.D. Bernal,[12] who was working at London University at that time. Bernal is regarded as the father of the application of X-ray crystallography to molecular biology. He had been kicked out of Cambridge in 1937 by Ernest Rutherford, which probably saved him from the temptation to get too close to the infamous Cambridge communist spy ring, led by Kim Philby. Several Nobel Prize winners were nurtured in his Cambridge laboratory, including Dorothy Hodgkin and Max Perutz. He continued to spawn Nobel Prize winners in London, including Aaron Klug and the famous non-awardee, Rosalind Franklin, whose X-ray structural analysis of DNA was poached by Crick and Watson in 1953. Bernal himself was never awarded the Nobel Prize even though he had worked out the structure of oestrin, cholesterol, vitamins B_1 and D_2, pepsin, the sterols, and the tobacco mosaic virus. The Stockholm Nobel Committee is notoriously conservative, and the history of science in the 20th century is littered with the names of sages like Bernal who were deemed to be too flaky[13] at the time to receive the prize.

In his 1949 lecture Bernal proposed that mineral surfaces were the ideal sites to organize these simple organic molecules to produce a putative self-replicating molecule. In parentheses, it should be noted that, like a good Stalinist, he also paid due respect to the contribution that Engels made to the theory of abiogenesis. Bernal was particularly taken with clay minerals, since these are quite complex mineral polymers in themselves and readily adsorb a large range of organic compounds. The secondary attraction of this process to Bernal was the problem of the concentration of organic compounds required to build and maintain a primitive self-replicating system. He thought the oceans and even Darwin's original warm little pond would dilute the organic concentrations too much. By contrast, adsorption on a mineral surface would provide an effective concentration mechanism. Bernal's ideas were taken up by the Glasgow chemist Alexander Graham Cairns-Smith in 1982 in an influential book on the mineral origin of life in which clays were promoted as primary templates.[14] Since that time, there has been much experimental work with clays that has produced some complex biomolecules although not, as yet, a self-replicating system recognizable as an immediate precursor to life.

Further on in his lecture, Bernal discusses the energetics of the process and notes (p. 610) that

The existence of labile inorganic reactions is obviously of the most critical importance. Two of these, on account of their extreme abundance, are known to be specially important—the oxidation and reduction of ferrous and ferric iron, and of sulfhydryl,—SH, to disulfide—S-S-. Early life, in the absence of atmospheric oxygen must have proceeded almost entirely by the enzymatic utilization of these transformations.

Of course we are familiar with these transformations: in particular the transformation of –SH to disulfide is what we see in the formation of pyrite, an iron disulfide, from hydrogen sulfide; and the transformation of ferric to ferrous iron is what we see when ferric oxides react with hydrogen sulfide to form pyrite. Gunther Wächtershäuser, a German chemist–turned–patent lawyer, independently latched on to this idea in 1990[15] and proposed that pyrite would provide an ideal mineral substrate for transforming simple prebiotic molecules to more complex self-replicating ones: pyrite was not only abundant on the early Earth but it also provided a potential energy source to drive the reaction (as had been intimated by Bernal) and included sulfur, an element essential to life and a necessary component of early organic molecules. I showed that this reaction had first been proposed by the father of chemistry, Jöns-Jakob Berzelius, in 1845 but had subsequently been quietly forgotten. The energetics are attractive since the reaction is exothermic and each molecule of pyrite that is formed produces around 30 kilojoules of chemical energy: if you remember that a single-bar, 1-kilowatt electric fire produces 1 kilo-joule per second, you can see the potential of this reaction. I described the kinetics and mechanism of this reaction and showed that it occurred quite rapidly at lower temperatures.[16] As a byproduct of this study, we demonstrated the then-shocking fact that hydrogen sulfide was quite a good oxidizing agent. This reflects the discussions in Chapter 7 where we reviewed the fact that oxygen is not the only oxidizing agent: anything can do this as long as it can capture electrons.

This energy bonus and the presence of sulfur made the idea of the formation of pyrite as a key catalyst in the formation of the first self-replicating organic molecules very attractive. The result of this process would be a two-dimensional, self-replicating molecule on the pyrite surface. The Wächtershäuser hypothesis is often thought of as *metabolism first,* since it suggested that a cycle of chemical reactions occurs first, which then produces energy that other processes can latch onto. This would enable even more complex compounds to be produced. Thus metabolism was established before genetics. The Wächtershäuser hypothesis can be tested experimentally, and a large number of experiments have been carried out since 1990. By 2000, Wächtershäuser was able to demonstrate a complete metabolic cycle, starting with inorganic carbon monoxide and ending with peptides, the basic molecules of proteins, all driven by pyrite.

Since this early work—and because these ideas are experimentally testable—there has been a substantial increase in our knowledge about the reactions involving iron sulfides in general in the synthesis of prebiotic molecules. In 1997, the Glasgow University ore geologist Mike Russell proposed that life began in iron sulfides in deep-ocean hydrothermal vents. The original idea here was to solve the problem of the formation of cells. All modern life occurs in cells, and these protect the biochemical reactions from variations in the environment so that the reactions can be repeated time and time again. The cell is a complicated object, as it needs to maintain stable conditions while still connecting to the outside world to import metabolites and export waste products. Another impasse seemed to be blocking the way to abiogenesis: the biochemistry needed a cell to be maintained and the cell needed the biochemistry to be formed. Russell's insight was that iron sulfides in the deep-ocean vent deposits were not compacted and the spaces between the crystals could provide the original cellular space, creating a stable environment and a surface on which the organic molecules could absorb. In this case, the energy to power the prebiotic syntheses comes not from pyrite formation but through the chemical gradients established by the iron sulfide walls of the inorganic proto-cell. With Michel Filtness, a senior honors student, we tested this idea in 2003 and measured the chemical gradients across a simple iron sulfide barrier and showed it is possible to separate two different solutions this way.[17] The problem is with the stability of the barrier: with two different chemical solutions on either side, it tends to migrate into the less acidic one.

One of the attractions of the hypothesis of the involvement of iron sulfides like pyrite in the origins of life is that, today, iron-sulfur proteins play dominant roles in most major metabolic cycles and are present in all life forms. This suggests that the formation of these iron-sulfur proteins was at the root of the creation of life as we know it today. They help transport electrons around organisms. With George Luther, we identified free-standing clusters of iron sulfide molecules in aqueous solutions both in the laboratory and in the environment and defined their biochemical significance.[18] We showed that these had the same structure as the iron sulfides at the heart of the iron-sulfur proteins, which are present in all life forms on Earth at present. This brought me full circle: Max Perutz, the Nobel Prize winner, had contacted me in the late 1960s about the synthesis, structure, and chemistry of the mineral iron sulfides I had made as a graduate student. Perutz had found similar molecular arrangements in the iron sulfides he discovered in the centers of iron-sulfur proteins. These iron-sulfide proteins are central to all life processes since they transmit electrons, allowing organisms like us and sulfate bacteria to obtain energy from chemical reactions. I still have his handwritten letter somewhere. He was off to give a lecture at a meeting in Spain and wanted the latest data—and there were no e-mails in those days. In 1972 Richard Holm and his group at Harvard University synthesized analogs of these iron-sulfur proteins.[19]

Over the past decade these ideas have morphed and become more sophisticated as further discoveries have been made in the natural world and in the test tube. The discovery of huge, white, deep-sea edifices made of lime on the floor of the Atlantic Ocean in 2000 caused the Russell group to move the focus of their research to these environments.[20] Reactions between the seafloor rocks and seawater produced green serpentine as well as methane and hydrogen. The fluids are highly alkaline, so reactions with seawater would produce even greater chemical potential energy for the synthesis of biomolecules. They are also at a more amenable temperature for prebiotic syntheses than the black smoker vents: 40° to 90°C rather than 400°C of the deep-ocean sulfide vents. Iron sulfides still produced the proto-cells but were formed through the reaction between sulfide and dissolved iron in the iron-rich contemporary seawater.

In 2010 my graduate student Shanshan Huang made original contributions to the nickel catalysis showing that nickel sulfides catalyzed some important prebiotic organic syntheses.[21] Both the Wächtershäuser and Russell groups have further studied the effects of transition metal catalysts on prebiotic organic chemistry, and both have noted the importance of nickel in these processes. Wächtershäuser and his colleagues have focused on high temperatures (e.g., 160°C) for the origins of life and have synthesized a number of important biomolecules in this surprisingly high temperature range. They argue that, in fact, the higher temperatures enable the reactions to go faster.

In 2007 another of my graduate students, Bryan Hatton, showed that nanoparticulate and dissolved iron sulfides denatured nucleic acids; that is, they caused them to uncoil and thereby lose their function.[22] This was worrying to me in the context of an iron-sulfide–based origin of life hypothesis since it seemed to suggest that iron sulfide and nucleic acids were incompatible. But Bryan assured me that this was really a good thing: it provided the possibility of rapidly producing multiple nucleic acid architectures with the consequence that the one that worked would be formed faster.

The hypothesis that iron sulfides were involved in the origin of life has become known as the *iron-sulfur world* hypothesis. It has taken on various forms, but it basically reflects the enhanced role that iron sulfides like pyrite played in an early Earth environment that had no oxygen around to oxidize them to rust. It closely links the sulfur contents in biological molecules, the possibility of coupling the energetics of iron sulfide reactions to organic syntheses, and the role of mineral surfaces in concentrating organic compounds and organizing them.

Pyrite and the Earliest Organisms

As described in Chapter 8, sulfate-reducing bacteria separate the various isotopes of sulfur in a characteristic way when they reduce sulfate to sulfide. This characteristic signature is then preserved in pyrite. Since pyrite is a fairly resistant mineral, this biological signature can be recognized in pyrite throughout geologic time. We mentioned before that the 3.85-billion-year-old Greenland pyrite does not show this biological signature. By contrast, 3.55-billion-year-old pyrites from Western Australia do show it. We therefore conclude that biological sulfate reduction—or, more strictly, a biological process that produced the isotopic signature in the pyrite—had evolved by 3.55 billion years and was probably not around at 3.85 billion years ago.

The well-preserved 3.5-billion-year-old sedimentary rocks from the Pilbara region on northwest Australia contain the oldest fossils on Earth. These are the remains of tiny microorganisms preserved in silica. I have some bits of these 3,500-million-year-old rocks on my desk. They vary from a rather unprepossessing gray rock with white silica fragments to a pretty red-and-white banded silica rock.

We can call the idea that the same isotopic characteristics imply a similar biochemical process the Schidlowski Principle after Manfred Schidlowski, the German geochemist who first proposed this in 1988 for carbon isotopes.[23] The argument is based on the idea that other processes give different isotopic signatures, so it would be a coincidence if similar ancient signatures did not have the same causes as today. There is not much evidence to support this principle, and it is essentially an example of the application of Ockham's Razor[24] to a problem that appeared intractable at the time. Since then, the use of all four stable isotopes of sulfur has provided more detailed insights into the biochemistry of sulfur isotope fractionation.

The biochemical process in the case of modern sulfate reduction is the interaction of a series of specific enzymes or active proteins to reduce sulfate to sulfide and to couple this to metabolism. The proteins involved are produced through the coding of sets of bases, called genes, on the DNA of the organisms. A cluster of just seventeen genes is involved in modern sulfate reduction.[25] Looking at the sporadic distribution of this gene cluster throughout all of today's life forms, they appear irregularly in a variety of single-celled microorganisms, many of which are not even remotely related. By analogy with macroscopic modern life, the same cluster of genes are found in organisms belonging to fundamentally even more diverse groupings than animals and plants. Their random distribution means that the cluster of sulfate-reducing genes has been transferred into the DNA of other cells by processes akin to infection rather than by an evolutionary or hereditary process. Where the gene cluster has settled successfully in the new DNA, the organism has thrived and is today living happily in some oxygen-free mud or even high-temperature volcanic vent, reducing sulfate to sulfide. Thus the isotopic signatures of ancient pyrites do not necessarily prove sulfate reduction by any particular life form, such as bacteria. The earliest organisms responsible for providing these sulfur isotope signatures were probably single-celled microorganisms, but they may not have been bacteria at all. They may have been some sort of primitive organism that is long extinct that possessed this particular gene cluster.

One of the key enzymes coded for in the modern sulfate-reducing gene cluster is shown in Figure 9.1. I have a particular affection for this enzyme since I first came across it when I worked with John Postgate's group at the National Physical Laboratory, where this substance was first isolated. John Postgate was the doyen of sulfate-reducing microbiologists, and it was his work in the mid-20th century that originally defined the characteristics of these microorganisms and their classification. He extracted a bright-green substance from these organisms, which he called *desulphoviridin*, from *desulpho*—relating to sulfate reduction—*viridis*, Latin for "green," and described it as a pigment. We did not know exactly what it

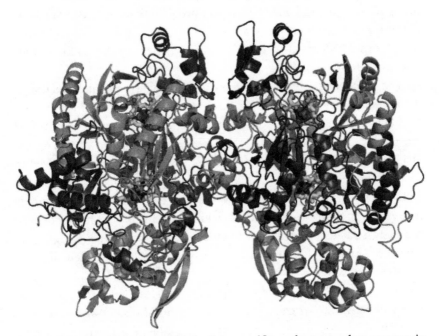

FIGURE 9.1. The structure of dissimilatory sulfite reductase, a key enzyme in biological sulfate reduction (see color plate). From T.F. Oliveira et al. 2008. The crystal structure of *Desulfovibrio vulgaris* dissimilatory sulfite reductase bound to DsrC provides novel insights into the mechanism of sulfate respiration. *Journal of Biological Chemistry*, 283:34141–34149.

did but initially used it to characterize sulfate-reducing bacteria. It was exceptionally abundant, and the organisms seemed to put a lot of effort into making it, so it had to play a key role in their metabolism. Later it was shown to be an enzyme called *dissimilatory sulfite reductase,* and it is central to the biochemical process by which bacteria reduced sulfate to produce hydrogen sulfide.

Figure 9.1 demonstrates the extraordinary complexity of these biomolecules. This one contains about 10,000 atoms of carbon, hydrogen, oxygen, nitrogen, phosphorus, and sulfur. And remember this is just one of the enzymes that are coded for by the cluster of sulfate-reducing genes. Like DNA, this enzyme is also the product of over 3.85 billion years of evolution, so the sulfate-reducing enzymes in the earliest organisms may not have been as complex or as efficient as this. This enzyme is interesting since at its heart are iron sulfide molecules, which might be expected in a system in which iron sulfides like pyrite are formed. It is also remarkable that this enzyme can work the other way: catalyzing the oxidation of sulfide to sulfur and sulfate. It occurs in microorganisms that earn their living from the oxidation of pyrite, causing acid mine water, for example, as described in Chapter 7. This oxidation process did not really take off biologically until there was free oxygen in the atmosphere around 2 billion years ago. It is an attractive idea that Mother Nature made use of what was around in her enzyme factory at the time to combat the new poison, oxygen, which was contaminating the contemporary atmosphere. One mutation of a sulfate-reducing organism managed to switch its genes to make use of the sulfide rather than the sulfate, and Darwinian evolution took over from that point. In fact, the process was likely to have been more complex and gradual since, even today, there are microorganisms that have the apparatus to oxidize sulfide in an anoxic environment. Our original green extracts of desulphoviridin fluoresced red in ultraviolet light. Thus, more subtly, there appears to be some biochemical affinity between this enzyme and the photosynthetic pigments of some organisms, and this may have provided an alternative route to successful biological sulfide oxidation.

This enzyme operates in a key stage of the biochemical workshop that reduces sulfate to hydrogen sulfide in sulfate-reducing bacteria. It is consequently involved in producing a large part of the isotopic signature that characterizes biologically produced sulfur in pyrite. This suggests that this molecule, or a similarly functioning precursor, had therefore evolved in the period of around 300 million years between evidence for the earliest life on Earth and the first isotopic evidence for biologically sourced sulfur in pyrite. This enzyme sits at the core of the biochemistry that produces the large sulfur isotope fractionation that is preserved in these ancient pyrites. The evolution of complex biochemistries involving large numbers of biomolecules therefore happened relatively rapidly after the origin of life on Earth. It seems to point to an inevitable process as soon as the first self-replicating organic molecules were formed.

In 2011 David Wacey, a young research fellow at the University of Western Australia, described fossil microorganisms in 3.4-billion-year-old sandstones that contained nanoparticles of pyrite in their cell walls. The pyrite grains show the characteristic signatures of biological sulfate reduction and further confirm the suggestion that sulfate reduction had already evolved 3.4 billion years ago, near to the dawn of recorded life on Earth. The apparent development of this gene set or, more likely, a precursor set so early in life's journey is remarkable.

Pyrite, Biofilms, and Ancient Lagoons

The oldest pyrite-forming microorganisms—the 3.4-billion-year-old fossils from Western Australia described earlier—are associated with *biofilms*. A biofilm is a group of cells immobilized on a surface, such as sand grains, and frequently embedded in an organic polymer matrix of microbial origin. This polymer matrix is produced externally by some bacteria, often in enormous quantities. One of these strains of sulfate-reducing bacteria I first cultivated in the 1960s used to regularly fill up the test tube with this stuff under certain conditions. We did not know what it was and called it an organic *coacervate*—one of those terms scientists apply to things that they do not understand and that give the false impression of deep sagacity. The 3.4-billion-year-old biofilms formed on sand grains deposited on one of the Earth's earliest preserved shorelines. It might be noted here that *sand* in environmental terms refers to a grain size and not a material. Sands can be made of silica, which is quite sharp to lie on but typical of temperate beaches such as in Wales, or it can be made of carbonate, giving the softer, luxurious, shimmering white beaches of more tropical climes. The 3.4-billion-year-old beaches were made of silica and pyrite sand. There was no free oxygen in the atmosphere, and the pyrite eroded out from the rocks and was taken down the rivers to the sea just like the silica.

The discovery that the earliest sulfate-reducing microorganisms were associated with microcrystalline pyrite in biofilms 3.4 billion years ago brings us back to the framboids, those tiny, raspberry-like pyrite spheres discussed in Chapter 4. These forms of pyrite are extremely abundant in sediments, and a thousand billion are being formed every second. They are remarkable because of the extraordinary organization of the thousands of pyrite crystals that make them up. There is an obvious parallelism between the form of these framboids and some forms displayed by bacteria and microorganisms. In 1923 the German ore geologist Hans Schneiderhöhn studied framboids in the 200-million-year-old, copper-rich shales of Poland and concluded that they were fossilized sulfur bacteria, with each spherule being the remains of a single bacterium.[26]

In 1957 the English micropalaeontologist Leonard Love dissolved the pyrite in framboids with nitric acid, leaving a honeycomb of organic matter (Figure 9.2). He thought that these were microfossils and named them *Pyritosphaera barbaria*, in honor of his wife Barbara. Jack Vallentyne, a

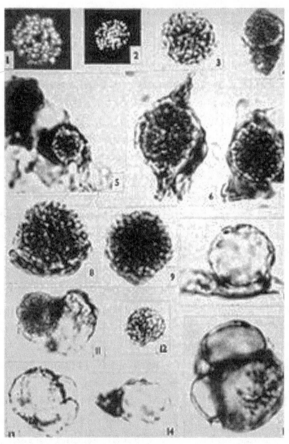

FIGURE 9.2. Original photomicrographs of the organic forms produced from dissolving pyrite framboids in nitric acid, which were termed *Pyritosphaera barbaria* and *Pyritosphaera polygona* by Leonard Love in 1957. From L. Love. 1957. Micro-organisms and the presence of syngenetic pyrite. *Quarterly Journal of the Geological Society, London*, 113:428–440.

US limnologist, found it irresistible to publish a paper titled *Concerning Love... by Vallentyne*, in which he cogently argued that framboids were not microfossils; he subsequently demonstrated this in a benchmark paper in 1963.[27] He showed that the organic residues were not present in most modern framboids. The microfossils soon became discredited species. However, the explanation of the presence of organic matter within some framboids was not resolved until the beginning of this millennium. The Schneiderhöhn work gave rise to a whole industry on framboidal pyrite as bacterial fossils, and Schneiderhöhn's fossilized bacteria still appear regularly, especially in the paleontological literature.

The problem was solved in 2001 by David Large, a chemical engineer from Nottingham University, UK, who showed that the organized organic residues resulted from growth of framboids within a biofilm. Large's work emphasizes the ubiquity of sulfate-reducing bacteria and the fact that they are all around us: he isolated his organisms from canals near Birmingham, England. The framboids grow in the biofilm, and, when the pyrite is dissolved, an organic honeycomb is left, appearing like a mass of tiny cells. However, the number and shape of the cells have been produced by the pyrite crystals and they are not individual bacterial cells.

The oldest sedimentary pyrite certainly identified is in the Pilbara region in northwest Australia. These pyrites are found in the lime muds of 3,500-million-year-old lagoons that also contain fossils of the oldest organisms on Earth. The characteristic biological structures in these ancient rocks are small, mound-like forms with thin crenulated internal layers called *stromatolites*. These are rare on the Earth today, but similar forms occur in the Hamelin Pool in Sharks Bay on the coast of Western Australia. These extraordinary structures are shown in Figure 9.3 together with their 3.5-billion-year-old ancestors. The modern stromatolites are produced by the action of cyanobacteria, an ancient group that produce organic matter from carbon dioxide using photosynthesis and release molecular oxygen. The ancestors of the cyanobacteria may have been responsible for the earliest free molecular oxygen in the Earth's atmosphere more than 2 billion years ago. They are very efficient, and you can see the oxygen, which is not very soluble, rising up through the water in chains of small bubbles. The reverse reaction at night produces carbon dioxide that reacts with calcium in the seawater to precipitate calcium carbonate or lime. The grains of calcium carbonate are trapped by the bacterial filaments and build up the layered mounds. The problem with the 3,500-million-year analogs is philosophic: there was no oxygen in the Earth's atmosphere 3,500 million

FIGURE 9.3. a. Modern stromatolites in the Hamelin Pool, Western Australia. b. Similar 3.47 billion-year-old stromatolites from Western Australia (see color plate). From D. Rickard. 2012. *Sulfidic sediments and sedimentary rocks* (Amsterdam: Elsevier, 801pp). Reprinted by permission.

years ago, so the ancient stromatolites are unlikely to have been formed by the same organisms that produce them today. Fossil stromatolites, which occur throughout the geologic column, do not include cellular fossils of the microorganisms that produced them. Looking at the ancient stromatolites in a microscope reveals a brownish mass of biofilm with no discernible cellular structures. By contrast, isotopic analyses of the carbonate reveal the same fractionation of the lighter isotopes that are characteristic of microbial processes.

When you visit the living stromatolites in the Hamelin Pool in Western Australia, the last thing you expect to find in this idyllic environment of pure-white carbonate sand and an azure-blue sea is foul-smelling black mud. But if you wade into the clear lagoonal waters between the stromatolites and disturb the topmost carbonate sediments, you will find a beautiful black sediment full of iron sulfide that produces the delightful smell of rotten eggs: the sulfate reducers are there working away in abundance. The sulfate-reducing bacteria are enjoying an almost ideal environment here: the lagoonal water is strongly enriched in sulfate since the bay has only limited access to the sea; evaporation is strong in the subtropical, restricted environment; and the cells of the cyanobacteria provide a plentiful supply of organic matter. So when I visited the 3,500-million-year-old Pilbara fossil stromatolites, I asked my guide, the Australian Geological Survey Regional Geologist for the Pilbara region Martin Van Kranendonk, why there was no pyrite to be seen. In fact, there were one or two iron oxide casts of what was perhaps originally pyrite. I looked at the rocks with a microscope and could find only a few specks of pyrite but not the rich layer of pyritic carbonate I would have expected by analogy with the modern Hamelin Pool stromatolites. Martin explained that weathering processes during the past few million years had silicified the surface layers of the Pilbara regolith. Drilling down a few meters beneath the silicified layers revealed a rich pyrite layer. Interestingly, the detailed analyses of the rarer sulfur isotopes of these pyrites proves that they have been produced by microbial processes. The sulfate in the Pilbara lagoons probably originated locally from volcanic sulfur dioxide. Biological sulfate reduction was ongoing 3,500 million years ago in this environment, even though it probably did not have the global significance that it has today.

If we were to transport our intrepid BBC radio reporter, whom we met in Chapter 8, back 3,500 million years to the Pilbara lagoon, he would not describe the idyllic lagoonal environment of today. First, he would be wearing a spacesuit and breathing oxygen from a cylinder strapped to his back, since there would be no oxygen in the atmosphere. The scene would be far from tranquil, since the bay waters were fed from the emissions of an active volcano. It is likely that the sky would be orange-colored rather than blue through the mixture of a cocktail of sulfurous gases and methane from the volcanoes. In his spacesuit the reporter would not be able to smell the air; this is probably a good thing since it would smell of rotten eggs and burnt sulfur and would probably be quite lethal to breathe. It would also be dangerous to expose skin to this mixture: hence the spacesuit. The sun would be 3.5 billion years younger than today, smaller and up to one-third less bright, so even on a sunny day it would be a bit gloomy. The water would not be blue, since this is reflecting the sky color today. It would likely be a darker color, possibly brown as the orange sky is reflected in a greener ocean with enhanced dissolved iron concentrations. It is even doubtful if the sands would be white, since the reason for that today is that the sulfate-reducing bacteria are forced away from the surface by the oxygen in the modern atmosphere. Three and half billion years ago, in the absence of atmospheric oxygen, the black iron sulfide would be at the sediment surface and even the overlying water would have been sulfidic. The dark-brown sea would lap onto a barren, rocky land surface with no hint of vegetation and no covering of soil.

It was in this strange world that life on Earth first evolved and that is recorded in the unique rocks of the Pilbara region. Although this environment is inimical to us and very unfamiliar, it would not be strange to much of the modern life that thrives beneath the surface of the planet and around the deep-ocean vents. This is one reason I describe this environment as one where early terrestrial life evolved rather than one in which it originated. The paradox of trying to define the environment for the origin of life is that all these environments appear to be present somewhere on the planet today. So either abiogenesis is occurring unbeknownst to us today and the new self-replicating organic molecules are being eagerly gobbled up by established life forms, or there was something peculiar about the early Earth environment that we have not yet discovered. Of course one way out of this paradox is to suggest that life originated on another planet or moon in the solar system and was brought to Earth on meteorites and asteroids early in Earth's history.

As I pointed out earlier, research into the origin of life is at present mainly laboratory based. It seems we can never know exactly when and where life on Earth originated; the best we can do is make it in the laboratory and extend these findings to the Earth system. Against this background, the idea that life began on another planet or moon merely extends the range of conditions available for the laboratory experimentation to cover the whole of the solar system. You may feel that this is a bit of a cop-out, but to the researcher it makes origin-of-life research one of the most exciting areas of current science. And in this pyrite may play a key role.

Notes

1. Aristotle. 350 BCE. *The History of Animals*, Book V, Part 31. Translated by A.L. Peck. 1989. *Aristotle History of Animals*. Loeb Classical Library (Cambridge, MA: Harvard University Press), 422pp. Anaximander's works are not preserved, and his ideas on spontaneous generation are mainly known through Aristotle.

2. L. Pasteur. 1862. Mémoire sur les corpuscles organisés qui existent dans l'atmosphère. Examen de la doctrine des générations spontanées. *Annales des sciences naturelles (partie zoologique) Sér.*4.16:5–98. Pasteur's famous swan-necked flask experiment disproved spontaneous generation. Pasteur sterilized identical broths by heating them in identical swan-necked flasks. The swan-neck was broken on one flask, allowing the ingress of microbes from the air, which immediately multiplied in the warm broth. The broth in the other flask remained sterile.

3. J. McPhee. 1981. *Basin and Range* (New York. Farrar, Straus and Giroux), 216pp.

4. The Earth, the solar system and everything was formed 4.57 billion years ago. However, it took 100 million years of impacts and planetary collisions for the Earth to finally reach its present mass. See A.N. Halliday. 2001. In the beginning... *Nature,* 409:144–145, for a nice summary of the schedule.

5. A.F. Trendall, W. Compston, D.R. Neson, J.R. de Laeter, and V.C. Bennet. 2004. SHRIMP zircon ages constraining the depositional chronology of the Hamersley Group, Western Australia. *Australian Journal of Earth Sciences,* 51:621–644.

6. P.E. Smith, N.M. Evensen, D. York, and S. Moorbath. 2005. Oldest reliable terrestrial ^{40}Ar-^{39}Ar age from pyrite crystals at Isua, west Greenland. *Geophysical Research Letters,* 32:L21318.

7. M.T. Rosing. 1999. C^{13}-depleted carbon microparticles in > 3700 Ma sea-floor sedimentary rocks from west Greenland. *Science,* 283:674–676.

8. In 1924 Oparin wrote a short paper titled *Proiskhozhdenie zhizny* (Origin of Life; Moscow: Izd. Moskovshii Rabochii), which he expanded into a book in 1936. In Anne Synge's translation: "These compounds (i.e. hydrocarbons) must also have arisen when carbides and steam met on the surface of the Earth. Of course some of these must immediately have been burnt, being oxidized by the oxygen of the air." Oparin goes on to argue that "during their stay in the hot wet atmosphere of the Earth, they (i.e. hydrocarbons) must certainly have combined with oxygen and given rise to the most varied substances."

   The first English translation of Oparin's 1936 book, *Origin of Life,* was published in 1938 (New York: Macmillan), 270pp. By this time, Oparin had, by his own admission, been persuaded by the investigations of the great Russian geochemist and mineralogist Vladimir Vernadsky, who first showed in 1926 that oxygen in the Earth's atmosphere was the product of biological processes. See V.I. Vernadsky. 1926. *Biosfera* (The Biosphere; Leningrad: Nauchn. Khim.-Techn. Izd-vo). French edition: Paris, 1929.

9. J.B.S. Haldane. 1929. The origin of life. In *Rationalist Annual* (London: Watts & Co.), pp. 148–169.

10. F. Engels. 1883. *Dialectics of Nature.* Translated by Clemens Dutt, 1940 (London: Lawrence and Wishart), 383pp. J.B.S. Haldane wrote the preface to this first English translation of the German original. The book is one of the pillars of dialectical materialism and required reading for Marxists.

11. The products of these and later experiments by Stanley Miller were reanalyzed using modern methods in 2011, and an even greater variety of more complex organic chemicals were found to have been produced. E.T. Parker, H.J. Cleaves, J.P. Dworkin, et al. 2011. Primordial syntheses of amines and amino acids in a 1958 Miller H_2S-rich spark discharge experiment. *Proceedings of the National Academy of Sciences,* 108:5526–5531.

12. J.D. Bernal. 1949. The physical basis of life. *Proceedings of the Physical Society, Section A*, 62:537–558.

13. One of Bernal's amazing faults was a blind acceptance of Lysenkoism, the official Stalinist doctrine of plant genetics well in to the 1950s, when it died with Stalin.

14. A.G. Cairns-Smith. 1985. *Genetic Takeover and the Mineral Origins of Life* (Cambridge, UK: Cambridge University Press), 477pp.

15. G. Wächtershäuser. 1990. Evolution of the first metabolic cycles. *Proceedings of the National Academy of Sciences*, 87:200–204.

16. D. Rickard. 1997. Kinetics of pyrite formation by the H_2S oxidation of iron(II) monosulfide in aqueous solutions between 25° C and 125° C: the rate equation, *Geochimica et Cosmochimica Acta*, 61:115–134.

17. M.J. Filtness, I.B. Butler, and D. Rickard. 2003. The origin of life: The properties of iron sulfide membranes. *Applied Earth Science (Transactions of the Institute of Mining and Metallurgy B)* 112:171–172.

18. G.W. Luther and D. Rickard. 2005. Metal sulfide cluster complexes and their biogeochemical importance in the environment. *Journal of Nanoparticle Research*, 7:713–733.

19. T. Herskovitz, B.A. Averill, R.H. Holm, J.A. Ibers, W.D. Phillips, and J.F. Weiher. 1972. Structure and properties of a synthetic analogue of bacterial iron-sulfur proteins. *Proceedings of the National Academy of Sciences*, 69:2437–2441.

20. For some of the latest morphs of the hypothesis, see M.J. Russell, W. Nitschke, and E. Branscomb. 2013. The inevitable journey to being. *Philosophical Transactions of the Royal Society B*, 368:20120254—but this is not for the faint-hearted.

21. S-S. Huang. 2010. *Nanoparticulate nickel sulfide* (PhD thesis, Cardiff University), 175pp.

22. B. Hatton. 2007. *The emergence of nucleic acids in an iron-sulfur world* (PhD thesis, Cardiff University), 194pp.

23. M. Schidlowski. 1988. A 3800 million year record of life from carbon in sedimentary rocks. *Nature*, 333:313–318.

24. Ockham's Razor is the popular name for the philosophical principle announced by the English Franciscan friar William of Ockham in the 14th century. It states that where there are competing hypotheses, the hypothesis with the fewest assumptions should be selected.

25. M. Mussmann, M. Richter, T. Lombardot, et al. 2005. Clustered genes related to sulfate respiration in uncultured prokaryotes support the theory of their concomitant horizontal transfer. *Journal of Bacteriology*, 187:7126–7137.

26. H. Schneiderhöhn. 1923. Chalkographische Untersuchung des Mansfelder Kupferschiefers. *Neues Jahrbuch für Mineralogie, Geologie und Palaontologie*, 157:1–38. S.V. Berg. 1928. Fossilifierade svavelbakterier uti alunskiffern på Kinnekulle.

Geologiska Föreningen i Stockholms Förhandlinger, 59:413–418, also suggested that the pyrite microspherules in the Swedish Alum Shale were fossilized sulfur bacteria.

27. J.R. Vallentyne. 1962. Concerning Love, microfossils and pyrite spherules. *Transactions of the New York Academy of Science, Series 2,* 25:177–189. J.R. Vallentyne. 1963. Isolation of pyrite spherules from recent sediments. *Limnology and Oceanography,* 8:16–29.

Full Circle

THE THESIS IN this book is that pyrite has been a key material in the development of our civilization and culture. It has figured in the foundation of nations and key industries, in the development of science, and in our current understanding of the nature of matter. It has played a key role in the development of our culture mostly through its use in the most important of human inventions: the taming of fire. Pyrite has determined the nature of the Earth's surface environment and the origin and evolution of life itself.

I have discussed how pyrite affects our present environment through its key role in the great biogeochemical cycles of fundamental substances like oxygen and carbon and how this has continued through over 4,000 million years of Earth history. This long history of the centrality of pyrite to the Earth system has enabled confident predictions about how pyrite is going to affect future Earth environment through acidification of atmospheres, rivers, and soils and eutrophication of the oceans, for example.

Pyrite has played a central role in the development of humankind for the entire 200,000 years of the existence of *Homo sapiens sapiens*[1] and this is unlikely to end now. It seems incontrovertible that pyrite will play a similar role in future human development as it has for the last 200,000 years. In this chapter I return to some of the themes from previous chapters and show how pyrite is still influencing our society and how this is likely to continue into the future.

Pyrite, Alchemy, and Metal Extraction

Gold occurs naturally in two basic forms: visible gold and invisible gold. *Invisible gold* was a term used by one of the greatest of 20th-century gold prospectors, John Livermore, to describe gold that "would not pan"—that is, gold that did not appear in the prospector's pan when the crushed rock or natural gravel was gently swirled around with water. Livermore discovered invisible gold in Nevada, which led to the 1980s gold rush in that state that has, to date, produced gold to a value of over US$85 billion.

A microscopic image of invisible gold is given in Figure 10.1, which shows gold occurring as tiny blebs entirely enclosed within a pyrite grain. The reasons pyrite is often associated with gold are quite complicated. The solutions in the Earth that transport iron and sulfur to form pyrite are also likely to transport other metals, including gold. Recall the introductory sections of Chapter 7 where oxidation and reduction were discussed. In particular, Figure 7.1 showed that the sulfur in pyrite has one more electron than the sulfur in H_2S, so it is a bit more oxidized. Pyrite is slightly oxidized relative to other metal sulfide minerals, and this is one reason why it is so abundant on Earth: it straddles the surface zone between deep anoxic Earth and the oxygenated atmosphere. The slightly more oxidized

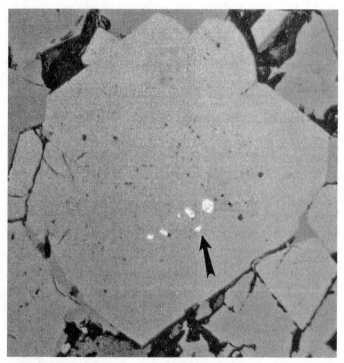

FIGURE 10.1. Hidden gold. Small blebs of gold (arrowed) within a 0.3-mm pyrite crystal. The photomicrograph was taken with a blue filter to show the gold more clearly against the normally brighter pyrite grain (see color plate).

environment where pyrite precipitates also destroys the sulfide complexes that keep the gold in solution and the gold precipitates as a metal. For these reasons, most gold deposits in the world contain pyrite as a more or less abundant mineral. In the case of invisible gold, tiny precipitated gold particles have been trapped in the growing crystal of pyrite. The amount of gold within the grain shown in Figure 10.1 is probably around 1% by weight, since gold is about four times as heavy as pyrite. A ton of this pyrite would then contain around 10,000 g of gold with a present-day value of over US$400,000. It is worth mining if the gold can be extracted from within the pyrite.

In 1783 the Swedish chemist hard-luck Scheele who we met in Chapter 4, discovered cyanide and was the first to demonstrate that gold would dissolve in cyanide solutions. The cyanide process was commercialized by John Stewart MacArthur, a Scottish chemist, and the brothers Robert and William Forest, both doctors of medicine, in Glasgow. Cyanidation is the most efficient way to recover gold and is used throughout the gold-fields of the world today. More than 90% of gold production is based on the cyanide process. In some low-grade ores, such as the Nevada deposits discovered by John Livermore, where there is not enough gold to pay for expensive ore preparation processes, the crushed rock is piled into heaps and sprayed with a dilute cyanide solution. After about six months the liquor exiting the heap carries sufficient gold in solution to be collectable. Of course, cyanide is highly poisonous, and although efforts are made to contain and denature it, accidents and escapes continue to occur.

In most modern gold ores, the gold is invisible and cannot be seen with the naked eye: it can only be observed with a microscope. In many years of working at our gold mine in west Wales, my colleague Alwyn Annels, the geologic expert on this deposit and a mining geologist of world renown, has only ever seen two specks of visible free gold.[2] In some gold-rich pyrite deposits, the gold cannot be seen even with a microscope. This is truly invisible gold. In these cases the gold occurs as clusters of atoms in the pyrite crystals and can be removed only by roasting the pyrite before cyanidation. This form of invisible gold can reach surprisingly high concentrations, and over 0.1% gold has been reported.

We can see how the old alchemists' idea that pyrite could change to gold came about. Some pyrite contains invisible gold. When this is burned in a crucible, the microscopic gold is exposed, but the particles are still too tiny to be seen with the naked eye among the detritus of iron oxides and sulfur. If you add mercury to this mixture, the mercury amalgamates with the gold and, being very dense, settles as a pool in the center of the crucible so that the detritus can be separated. You heat the crucible to drive off the mercury, and you are left with a small amount of gold. It appears as though the pyrite has changed to gold.

When Diego Delgado reported back to King Felipe II in 1556 after his exploration of the Rio Tinto in Spain (Chapter 7), he had some information he did not want to commit to a written message. It was so sensitive that he thought he should tell the king personally. We do not know this information, of course, but it was thought to be Delgado's observation that, if iron metal were put into the Rio Tinto and left a few weeks, it changed to copper. This process of extracting copper from acid mine waters such as the Rio Tinto had been carried out at least since Roman times. This was the same process the youngsters in Jerome, Arizona, carried out in the Verde River in the late 20th century, as described in Chapter 7. The secret thought of Delgado was probably that if iron could change to copper, could it change to gold? The apparent transmutation of iron into copper was ancient knowledge and formed a basis for the alchemists' claims about the transmutation of the elements. Add to this the parlor trick of extracting gold from pyrite, and you have the groundwork for the alchemists' theories. After all, you cannot fool all the people all the time and it is unlikely that all the people would have been fooled by the alchemists for the 1,500 years between Zozimos's original speculations in the 4th century CE and the dying breaths of alchemy in the 17th century without some supporting evidence.

There is no doubt that alchemy was the origin of the modern science of chemistry. The alchemists described what they saw in terms of the knowledge and culture of their time. What we now know as chemical reactions were processes that they saw as one substance changing into another. It is unfortunate that the English translation of this change was *transmutation*: the idea that one material *transmutes* into another rather than simply reacts or changes has a ring of mystery as well as apparent erudition. The mystery of change permeated the thinking of the ancient world from the *I Ching* through Ovid's *Metamorphoses* to the mystery of transubstantiation in the Catholic Church. The alchemists cannot be blamed for extending their observations of natural processes to more and more outlandish theories. This still happens today. As Mark Twain wrote,[3] "There is something fascinating about science. One gets such wholesale returns of conjecture out of such a trifling investment of fact."

Some of the most persuasive and remarkable changes in substances were where common iron changed into noble copper and humble pyrite could change into immaculate gold. Both processes involved pyrite. In this sense, pyrite can be said to have been at the heart of the development of chemistry as a science.

The ancient knowledge that one could put iron metal into an acid river like the Rio Tinto and pull out copper has been turned into a major industry in modern times. Chapter 8 was full of gloom and doom as the global implications of the production of acid through the burning and weathering of pyritic rocks, such as coal, were considered. But as with all dark clouds there is usually a silver lining. In this case it is the use of the acid produced through pyrite oxidation for the extraction of metals, such as copper, from ores that would otherwise be too low grade to treat. Some of the biggest ore deposits in the world are copper ores, called porphyry copper deposits, where the ore contains only a small amount of copper in a porphyry, a type of granite. For example, the Escondida deposit in Chile contains almost 4 billion tons of ore grading between 0.2% and 1% copper. You can see that even present production rates of 80 million tons of ore per year (i.e., 1 million tons of ore and waste rock shifted every day) could last fifty years. Most of the ore at Escondida contains sufficient copper for it to be extracted by conventional chemical methods. However, conventional treatment of the marginal ore with grades less than 0.7% copper is not profitable. This reserve, amounting to 1.5 billion tons or around 20% of the total copper ore, is piled into huge heaps of crushed rock and the copper is leached out by sulfuric acid. The first copper ingots from this process were produced in 2006, and currently some 20 million tons of marginal ore are being leached in this fashion every year, producing 18,000 tons of copper annually. The process extracts about 35% of the total copper out of the rock. The neat thing about all this is that the copper that is extracted by this process is virtually cost free, since the mining costs are covered by the conventional extraction of the richer ore.

The heap leaching process depends on the bacteria that oxidize pyrite, described in Chapter 7. These bacteria produce sulfuric acid and the oxidized, ferric form of iron that solubilizes the copper minerals. The bacteria do not attack the copper minerals directly, and hence the process depends on pyrite being present in the ore. In a typical heap leaching operation, the top of the pile of crushed ore is sprayed with water, which percolates down to the bottom of the heap. Here it is collected, sent through a metal recovery system, and then recycled to the top of the heap. This does not waste the sulfuric acid and dissolved ferric iron in the leach solution, and also maintains many of the bacteria. Various methods have been applied to increase the amount of oxygen in the leaching solution (since the bacteria need oxygen) and the concentration of carbon dioxide (since many of the bacteria in the microbial consortium convert carbon dioxide to organic matter). The introduction of specific strains of bacteria with high process rates has been tried but has generally been unsuccessful and the natural microbial flora is usually used. This is a highly complex synergistic consortium of diverse microorganisms that vary in composition throughout the heap and may vary with time. The optimization of the heap microbiota is still a bit of an art, and much research is being done in this area.

This is not some isolated process in a country far away but one that affects you today as you sit reading this book. Some part of the copper in the wiring that is bringing electricity to your house or office, powering the lights that enable you to read this book after dark or the computer that carries images of this book's pages, is probably sourced from this process. Pyrite is still at the heart of our everyday existence.

As we discussed in Chapter 2, pyrite may contain or be associated with a variety of valuable metals. However, apart from copper, bioleaching has been successfully applied only to cobalt, gold, and uranium recovery. The treatments of the ores of these three metals are distinct. Cobalt is released from pyrite, invisible gold is exposed in pyrite for chemical leaching, and uranium minerals are dissolved from the sulfuric acid produced from associated pyrite in the ores. The common thread in each of the extraction processes is pyrite and the microorganisms that increase the rate of pyrite oxidation by a million times compared to the abiogenic chemical process.

In the case of cobalt, the target mineral is cobaltian pyrite. This is collected from pyritic stockpiles at the Kasese mine in Uganda,[4] and the cobalt is leached out in bioreactors, large tanks of pyrite slurry with microorganisms busily oxidizing the pyrite to produce sulfuric acid and ferric iron and releasing the cobalt into solution. The tank system has some advantages over the heap process since the conditions in the bioreactors can be more closely controlled. For example, the temperature and the oxygen supply can be optimized for the bacteria. As is usual, the economic viability of the cobalt bioleaching plant at Kasese is highly dependent on the price of cobalt: between 2002 and 2004 the plant was put on care and maintenance as the cobalt price dipped below the economic cut-off price of US$12 per pound of cobalt. When the world cobalt price steadied around US$20 per pound in 2004, the plant was switched on again. If nothing else, this demonstrates the flexibility of this technology.

One of the problems with gold extraction is that much of the gold is included within pyrite and related minerals, as shown in Figure 10.1. This means that chemical leaches, such as cyanide, cannot get at the gold because it is protected by the pyrite. In the past the only way to release the gold was to roast the pyrite at 700°C or to digest it in strong acid in an autoclave at elevated temperatures and pressures. Both of these processes are very expensive. In 1986 industrial-scale testing of bio-oxidation of refractory gold ores was introduced by Gencor in South Africa. In this process the crushed ores are initially subjected to the attention of pyrite-oxidizing microorganisms in tank fermenters. The process exposes more of the gold, and subsequent cyanidation of the bio-oxidized concentrates can result in increases in gold recovery from just 30% to over 95%. The use of microorganic reactors to prepare pyritic gold ores for cyanidation has become widespread, and numerous mines in South Africa, Australia, and North America have various operating systems.

Uranium was produced in 1988 by bioleaching at the Dennison mine in Canada. Some 300 tons of uranium were produced with a contemporary value of US$300 million. By contrast with the heap leaching used for copper and the tank leaching used for cobalt and gold, uranium leaching took place in situ in the mine. The walls of the stopes, the underground rooms where the ore is mined, were sprayed with the acid mine drainage and the resulting uranium-rich solution was collected and the uranium extracted. The microbiological process is the same as described previously where microorganisms oxidize pyrite associated with the uranium minerals to produce the sulfuric acid and ferric iron solutions that leach the ore. The essential role of pyrite in this process means that only uranium ores with a significant pyrite content, such as those in eastern Canada, are amenable to bioleaching. This has limited the application of the technology worldwide.

The processing of these metals has come a long way since I first worked on them in the laboratory in 1966. I returned to the scene in the 1980s when Fred Pooley, a colleague in my department in Cardiff University, helped develop the Gencor process for the bioleaching of gold ores. The future of bioleaching of ores looks bright as ore grades become poorer and metal prices become higher, and there is a considerable amount of research going on at present into this technology. There is particular interest in the potential of real in situ bioleaching, where the ore deposit is not mined but fractured and the leach solutions are pumped down. The problem with this idea at present is the difficulty in recovering the solutions—they tend to disappear down fractures in the Earth's crust, never to be seen again. It seems that whatever the future of this technology, pyrite will be at the heart of it.

Pyrite and the Electronics Industry

Pyrite is a semiconductor; that is, it is neither a conductor like metal nor an insulator like most rocks. Metals are used in cooking pots, where heat needs to be conducted efficiently, and in wires, where electricity must be conducted. Rocks are used in building sensible homes. The old houses in our village in France are built of stone: stone is a good insulator so that the houses are cool in summer and warm in winter. The idea of pyrite as a good conductor sometimes and a good insulator at others echoes the problems of the ancients when they tried to classify pyrite. For example, the late-medieval mining geologist Agricola observed the contradictory properties of pyrite and classified it as a mixed species, somewhere between metal, stone, and sulfur (Chapter 5).

Semiconductors like pyrite can switch between being a good conductor or insulator under the effects of electric fields or light or by doping the material with traces of impurities. In pyrite, only a small amount of energy is required to release electrons from being chained to the atomic nuclei so that they can move freely in the material and conduct electricity. In other words, a small amount of energy will switch pyrite from behaving like an insulator to behaving like a conductor.

I first came across pyrite in this context when my brother and I built a cat's whisker radio in the early 1950s. The radio, or *wireless*, as it was called then, had a thin wire (the "cat's whisker") that touched a crystal and acted as a rectifier, and pyrite was one of the semiconducting crystals that could be used. The radio was called a *crystal set* because of the semiconductor crystal at the heart of it. This arrangement was successively succeeded commercially by valves and then solid-state devices, such as thyristors and transistors. However, the cat's whisker was the earliest semiconductor electronic device: it was the granddaddy of our current plethora of electronic gadgetry, and pyrite was at the start of it.

Satisfying the increased demand for electricity will be one of the fundamental problems faced by humankind over the next fifty years. At present, almost one-third of the world's population has no reliable electricity supply. It is estimated that over 30 million billion watts of extra power will be required by 2050, and supplying this by fossil-fuel generation not only is improbable but also will have a considerable impact on the Earth's climate. The obvious solution is to capture the energy from the sun using solar panels. However, as anyone who has installed solar panels on their homes will testify, current silicon-based solar panels are expensive. The energy cost, amortized over the 20-year lifetime of the panel, is around twice as much as wind- and natural gas–generated electricity. In order to contribute significantly to our electricity-generating requirements in the medium term, the industry will have to undergo a step-change in productivity and costs. This is where pyrite comes in as the most cost-efficient alternative solar panel material to conventional silicon.

Pyrite absorbs 100 times as much light as that of the present major solar cell material, silicon. A thin layer of pyrite, just 0.1 millionths of a meter in thickness, theoretically absorbs almost 90% of the solar radiation, whereas thicker current silicon-based systems harvest less than 20%. Silicon, although it is the second most abundant element in the Earth's crust, is expensive to extract. The cost of extraction is about US$1.7 per kilogram or over fifty times as much as pyrite. Pyrite is in fact more attractive in terms of cost and availability than all other natural photovoltaic materials.[5] Since only a very thin layer of pyrite is required to collect the sunlight, suspensions of tiny pyrite crystals, like the crystals that constitute the ubiquitous pyrite framboids of Chapter 4, might be mixed in a solvent and sprayed onto panels like paint. Considerable research is going on worldwide at present to synthesize pyrite crystals and films with various compositions in order to produce an optimal solar energy collector.

The other way to help resolve the world energy gap is to find a better way to store electricity. Electric automobiles are wonderful except for the fact that we are at present limited to a 100-mile working distance and a 24-hour charging cycle. Portable computers are fantastic—for eight hours until the battery runs out. We described in Chapter 2 how pyrite powers the global chemical industry through its role as a source material for sulfuric acid manufacture. Sulfuric acid is the most abundant manufactured chemical, and one use of it is in car batteries: it is the acid in the lead-acid battery. These lead-acid batteries are still used in automobiles, even though the technology is ancient, because they are rechargeable. As you drive around you recharge the battery. However, these lead-acid batteries are cumbersome and not suitable for many applications where a small solid-state battery is required. The problem with these small batteries is that they are not especially powerful or, in many cases, rechargeable. There have been many recent advances in battery technology. One of the most popular, and familiar to readers, is the development of lithium batteries. In the Energizer™ series of lithium batteries, lithium metal is the anode (the negative electrode), and pyrite is the cathode (the positive electrode). This pyrite is simple mined pyrite that has been ground down to 0.1-mm particles and stuck on aluminum foil in the battery. The battery works by a redox reaction (see Chapter 7) whereby the lithium metal is oxidized to produce lithium sulfide and the pyrite is reduced to iron. The redox reaction produces electrons, which we use as electricity. The lithium batteries are popular because they are relatively light, so the amount of energy per gram is optimized. At present these are not basically rechargeable, and the development of rechargeable lithium batteries is a major international target of technological research.

Pyrite is an attractive material for the electronics industry: it is widely distributed, cheap, and readily available. It has some environmental benefits in terms of the amount of energy required in transport and manufacture. All of these attributes are the same as those that originally placed pyrite at the core of early industrial development, as described in Chapter 2. It is interesting to speculate that the 21st century will see the burgeoning of a pyrite-driven electronics industry just as earlier periods witnessed the development of pyrite-driven chemical, pharmaceutical, and explosives industries.

The Pyrite Resolution Board

A few years ago I was sitting at my desk when the phone rang and a soft Irish voice asked me if I would be willing to act as an expert in a forthcoming court case in Dublin, Ireland. The job was a bit different from the expert witness role in US courts. Here I would sit with the judge on the bench and advise on the technicalities of the evidence. The court case was expected to last six months and concerned a class action by a group of Dublin householders against the builders of their homes. The householders were claiming that pyrite in the concrete used by the builders in the construction of their houses had caused their homes to break up, a sort of subsidence in reverse.

Nothing more happened for a time. Then several years later there was a spate of court cases in Ireland concerning this issue with builders, quarry owners, and insurers suing each other and appealing the court decisions in the usual fashion. In 2014, the High Court agreed to rule on 400 claims by residents against a number of large Irish companies. In the meantime, the worst-affected residents have no homes to live in: a group of families in north Dublin, for example, have been living in caravans for the past three years.[6]

In January 2014 the Irish government published the Pyrite Resolution Act, which aimed at "the remediation of private dwellings with significant damage caused by pyritic heave of hardcore under floor slabs." The Act concerned houses in Dublin and four counties in the Leinster province surrounding Dublin. The problem affects 74 housing estates in these areas with over 10,000 homes at risk of *pyrite heave*, in which the ground-floor slabs rise as pyrite oxidizes and the rocks swell, fracturing the walls and floors of the buildings. Remediation involves the removal of the pyritic hardcore beneath the concrete slabs in the foundations at an average cost of €45,000 per house, followed by the rebuilding costs. The Irish government has allocated €10 million to combat the problem.[7]

The problem is localized to this area because the local builders used a calcareous mudstone rock from the 350-million-year-old Tober Colleen formation, which forms a horseshoe around north Dublin, as backfill for houses in the region. This rock contains high pyrite concentrations, often in the form of Mother Nature's favorite texture (Chapter 4)—the framboid. It might be thought common sense that a soft, limy mudstone rich in pyrite might not be the best material to use for a level base for a concrete ground-floor slab. The problem is one we are now familiar with: exposing pyrite in crushed rock to the air and gentle Irish rain leads to the formation of sulfuric acid, with the help of an association of microorganisms. The sulfuric acid reacts with lime in the rock to form gypsum, which has twice the volume of the pyrite. The result is pyrite heave. The effect in Ireland is remarkably rapid and takes as little as two years to show itself. It is thought that the framboidal pyrite might be responsible for the rapidity of the effect in Ireland. In contrast to crystals of pyrite, which oxidize on the outside, the framboids may oxidize internally as the oxidizing solutions penetrate between the pyrite microcrystals.[8] The pyritic mudstone was known as a "non-premium" aggregate and was apparently used in Ireland to meet the exceptional demands for aggregate caused by the Irish property bubble around the turn of the millennium, which ultimately led in itself to the collapse of the Irish economy. The idea that the construction industry deliberately used this material because it was cheaper and that it could turn a fast buck has not been suggested. Details of the potential risk involved in using pyritic mudstone in building foundations were not well known at that time and certainly were not addressed in the training of engineers and architects.

The problem is not limited to Ireland, however. Aggregate of pyritic mine waste and shale in hard-core aggregate caused heave in ground-floor slabs in northern England in the 1970s. The problem is known in the southwestern United Kingdom as the *mundic* problem. *Mundic* is an old Cornish word for *pyrite* and was used to describe local copper ores. It is now used to describe the deterioration caused by using mine waste containing pyrite as aggregate in concrete mixing. Houses built in the first half of the 20th century are routinely checked for this problem in this area if they are constructed from concrete blocks. The extent of the problem is unknown, but some estimate that 15% of the properties built in the 50-year period before 1950 are potentially at risk.[9]

The problems of pyrite heave were originally identified in buildings built on bedrocks of pyritic shale rather than caused by the addition of pyritic backfill. The Rideau Health Center and Bell Canada Building in Ottawa suffered pyrite heave in the 1960s. One of my local hospitals at Llandough in the western suburbs of Cardiff suffered ground heave of up to 81 mm (over 3 inches) in 1982 through being built on pyritic shales. In Kentucky, pyrite oxidation in the Chattanooga shale has caused foundation problems in many buildings and roads. The Chattanooga shale is interesting in this context since it is a classic rock for pyrite framboids. It contains an unusual proportion of perfectly organized framboids, which caused great interest in the 1960s. In the context of pyrite heave, this abundance of framboids may be an additional reason why it may not make a good bedrock for construction. Occurrences of pyrite heave caused by construction on pyritiferous bedrocks have been reported in 18 US states as well as Norway, Sweden, and Japan.[10]

The Irish experience was exceptional. No one can guarantee that a builder in the future will not use mine waste of pyritic shale as backfill in construction. By contrast, the effects of pyritic bedrocks as foundations to building projects appear to me to be increasingly difficult to avoid as population increases and more construction activity results. The problem is similar to that discussed in Chapter 7: the construction activity exposes more pyritic rocks to the air and more pyrite is oxidized. The demand for land for construction as the population increases is accompanied by a greater risk that areas underlain by pyritic rocks will be increasingly used. The problem is not limited to buildings: pyrite heave has been observed in new freeways, dams, and even nuclear waste disposal sites.

The interesting thing about the Irish experience is that through the Pyrite Resolution Board, the Pyrite Act, the Pyrite Remediation Scheme, and the Pyrite Problem, pyrite has entered the public consciousness. For millennia, this mineral had been known under various names to the general public as a brassy yellow substance that looks like gold. There is no doubt that many competitors on any television quiz show—or its equivalent in ancient times—would know about pyrite. They would probably not know what it is made of or its importance to industry, science, culture, and civilization, but most of the general public would know the name and what it looked like. The Irish experience has changed this. Now pyrite—at least in the Dublin area—is the subject of newspaper headlines and television discussion programs. The increase in public awareness of this mineral and its role in everyday life and popular culture is likely to continue as pyritic rocks become more exposed to the atmosphere as a result of the increasing demands for building land for expanding populations with rapidly rising standards of living.

Pyrite, Free Radicals, and Life

I have already mentioned pyrite and DNA in Chapter 9. Here I discuss further how pyrite is helping us understand more about a particular biochemical process: the role of free radicals in biochemical reactions.

In this context, *free radical* does not refer to a special libertarian type of politician but to a type of atom or molecule with a particular arrangement of its electrons. I showed a solar system–like diagram of electrons in atoms in Figure 7.1 in Chapter 7. In fact this arrangement existed only as a representation of the real geometry of electrons in atoms for a few years in the beginning of the 20th century. Subsequently it was shown that electrons tend to be paired in their orbits. When an electron is missing from the outermost orbital, the remaining electron is unpaired. It is a truth universally acknowledged that an unpaired electron in an atom must be in want of another electron. These atomic and molecular species with unpaired electrons are called *free radicals*. These free radicals ravenously grab electrons from other species in order to satisfy their need for pairing. They are thus extremely reactive and do not exist for a long time. However, if they are continually produced then there is a significant standing concentration. In a typical free radical reaction the radical steals an electron from another species, which in turn becomes a free radical. This process continues so that a chain reaction is produced. Because they are so reactive, free radicals are very dangerous to our health: they snip the long chains of nucleic acid polymers like DNA.[11] The final result is disruption of the living cell. They have been implicated in aging and a number of human diseases, such as some cancers, cardiovascular disease, diabetes, and rheumatoid arthritis, and they are suspected as culpable in some neurodegenerative conditions such as Parkinson's and Alzheimer's disease.

Many of these free radicals are derived from oxygen. These oxygen free radicals are extremely dangerous and are the cause of the health advice for you to eat antioxidants, including five vegetables and fruits a day, which contain antioxidants that neutralize the oxygen free radicals. Since their lifetimes are so short, free radicals must be continuously generated to do substantial damage. And this is where pyrite comes in.

Pyrite reacts with water to produce oxygen free radicals.[12] It does this most effectively in the presence of molecular oxygen, but it also reacts steadily with water in the absence of oxygen to produce a significant supply of free radicals. These oxygen free radicals react almost nonspecifically with organic molecules, so any environment where pyrite is in contact with water will produce sufficient free radicals to degrade organic matter.

Oxygen free radicals created by the reaction of pyrite with water have also been implicated in the chemical oxidation mechanism of pyrite.[13] This reaction, which is catalyzed by microorganisms, is a key reaction in the Earth system, as discussed in Chapter 7. The free radical mechanism suggests that it is initiated as a chain reaction: the more that pyrite oxidizes, the more that can be oxidized. The oxygen free radicals produced during the reaction damage the microorganisms involved in the process, causing a decrease in the anoxic oxidation of ferrous to ferric iron.[14] This in turn has knock-on effects, since the ferric iron produced is a key component in oxidizing other compounds, as well as contributing to the overall rapidity of pyrite oxidation in acidic solutions.

Life has evolved many biochemical systems to defend itself against the ravages of free radicals. However, the development of an atmosphere containing molecular oxygen around 2.4 billion years ago must have been accompanied by the formation of oxygen free radicals, with consequent damage to organisms around at that time.[15] Indeed, the reaction of pyrite to this new oxygenated atmosphere would have increased the production of the deadly oxygen free radicals, analogous to the effects of the current enhanced exposure of buried pyrite through human activity (Chapter 7). It is quite possible that it led to a mass extinctions of microbial species comparable to the mass extinctions of macro-organisms during the past 500 million years (Chapter 8). The evolution of biochemical defenses against free radicals must have been a key aspect of evolution at that time.

One of the holy grails of free radical chemistry is trapping sulfide free radicals so they can be studied and their role in nature understood. These are known to exist in combination with organic molecules, but the discrete species in aqueous solutions are so reactive that they are extremely transient and difficult to trap and study. They are far more unstable than the analogous oxygen free radicals. The single sulfur atom of the disulfide pair in pyrite is a free radical (see Figure 7.1 in Chapter 7), since it has a lone electron in the outermost electron shell. However, it rapidly reacts with a sulfide atom to steal an electron to form a pair of sulfur atoms in the disulfide molecule. Indirect evidence for sulfide free radicals has been detected in aqueous solution through their effect on denaturing of DNA and other nucleic acid polymers (see endnote 11). There seems little doubt that sulfide free radicals will be collected and studied in the future, and they may play a key role in pyrite formation and its natural chemistry.

Extraterrestrial Pyrite

Although pyrite is the most common sulfide mineral on the Earth's surface, it is not the most common sulfide mineral in the solar system or, indeed, in the Earth as a whole. The most common sulfide is pyrrhotite and its related iron sulfide, troilite. These minerals have compositions approaching FeS in contrast to the FeS_2 of pyrite. FeS is analogous to H_2S where the sulfur only has one missing electron instead of two as in pyrite (see Figure 7.1 in Chapter 7): so FeS is more reduced than pyrite and pyrite is more oxidized than FeS. Pyrite dominates the Earth's surface environment primarily because it is stable in water at ambient Earth surface temperatures.[16] The region where FeS becomes stable relative to pyrite at these temperatures is one in which water itself starts to break down to its constituent components, hydrogen and oxygen.

FeS is not stable in water and oxidizes to form pyrite. However, the presence of pyrite is not necessarily indicative of water. FeS also reacts with both H_2S and sulfur to form pyrite in the absence of water at high temperatures. So we have to qualify the statement about pyrite and water: we can say that the presence of pyrite is indicative of water at lower temperatures, below around 100°C.

In the vacuum of space, H_2S and sulfur are not stable; one is a gas that is removed in a vacuum, and the vapor pressure of sulfur is so high that it sublimates in a vacuum. So the characteristic iron sulfide of most meteorites is a form of FeS, troilite, a mineral that is relatively rare on Earth. There are, however, some rare meteorites that contain pyrite. These are meteorites that originated on Mars. We know this because the gases trapped in the meteorites have the same peculiar composition as the Mars atmosphere, which has been sampled and analyzed by landers on the planet. These meteorites originated as fragments of Martian rock that were blasted into space as a result of meteorite impacts on the planet. These fragments then fell inward toward the sun under gravitational attraction, and at least 32 of them landed on Earth. They are difficult to find, and meteorite hunters mainly search for them in Antarctica, where the black stones stand out on the white snow and the slowly grinding glaciers bring up new meteorites every year. Several of these Martian meteorites contain a little pyrite, and this seems to be due to a high-temperature reaction between FeS and H_2S or sulfur.

One meteorite discovered in the Allan Hills of Antarctica shows pyrite in a fracture-filling with carbonate. This pyrite formed at lower temperatures and therefore suggests the presence of water on Mars when it was formed. The fracture-filling appears to have been a vein in the Martian crust analogous to the terrestrial veins so copiously described by the medieval mining geologist Agricola in Chapter 5. The sulfur isotope composition of the pyrite does not display a biological signature. The observation suggests that water was present in the Martian crust when the vein was formed, about 4 billion years ago, and points to a system similar to the terrestrial volcanically powered geothermal systems described in Chapter 5.

This meteorite attained worldwide notoriety when President Bill Clinton announced that it contained fossil microorganisms, the first evidence for life on Mars. In fact, these are more likely to be microscopic mineral inclusions, but the Clinton press conference served a purpose in getting Congress to allocate enough money to NASA to pay for its current Mars exploration program. The exploration program run by NASA's *Opportunity* rover discovered jarosite on the Martian surface. Jarosite is a hydrated sulfate of iron and potassium formed by the oxidation of pyrite. It is a common terrestrial mineral, and jarosite from Falun, Sweden, which resulted through oxidation of pyrite ore, is shown in Figure 5.5 in Chapter 5. This suggests that water was present on the surface of Mars, and it is concordant with the observation of hydrothermal pyrite in the Allan Hills meteorite. Water on the surface of Mars, in the form of large lakes, appears to have been present more than 3.8 billion years ago. Some scientists believe that there is evidence for more recent water in the form of occasional ice melts, but the Martian atmosphere, which is analogous to a freeze-drier on Earth, could have preserved the jarosite for billions of years, rather as if it had been set in aspic.

The Martian experience suggests that pyrite is going to be a key mineral in space exploration for similar reasons to its importance on Earth. Probing the mineral reveals information about the surface environment, including both the aqueous systems and the atmosphere, as well as preserving detailed signatures of biological, especially microbiological, processes. Iron is extremely abundant in the solar system, and sulfur is commonly found as well. Jupiter's moon Io, for example, is perhaps the most volcanically active body in the solar system, and its volcanoes are powered by sulfur. It is certain that as we learn more about the planets, moons, asteroids, and comets of our solar system, pyrite will play an important role. The shiny crystals of this humble mineral provide a record of the past, help explain the present environment, and allow predictions of what will happen in the future.

Not So Foolish Gold

Pyrite is well known as fool's gold. In Chapter 1 we traced the origin of this epithet to the 19th-century goldfields of the United States and showed how it had become a universal metaphor for something that appears valuable but actually has no or very little intrinsic worth. However, it turns out that pyrite does commonly contain gold, and modern metallurgy is able to extract the gold from pyrite and provide a considerable profit. Indeed, pyrite is probably the most important mineral hosting gold in the world today: most gold is produced from pyritic ores.

Many readers will be more familiar with natural free gold since, when visible to the naked eye, the grains are called nuggets. You are familiar with it, although it is unlikely that you have actually ever seen a real nugget. Most of the large lumps of free gold near the surface of the Earth have been found as humans have spread over the planet. However, there are still some places where large nuggets can be found near the surface.

Figure 10.2 shows a happy prospector with the nuggets of gold she had just found in the sandy soils of the Pilbara region in Western Australia. She is holding several hundred thousand dollars worth of gold in her hand. My wife, Simonne, was so impressed that the next day she collected a rucksack full of white quartz rock, which the prospector had told her contained the gold. By the end of the day she was exhausted from hauling this around and disappointed to be told that there was no gold in her rocks. The prospector had forgotten to mention that the type of white quartz rock that might contain gold is often stained red. This red stain is the result of oxidation of pyrite contained in the quartz and really just an indicator of the possibility of gold nuggets occurring buried in the soil nearby or settled in the gravels of a nearby dry river channel.

FIGURE 10.2. A happy prospector with a handful of gold (see color plate).

The process of the formation of gold nuggets is exactly as described earlier for invisible gold except that the gold particles have not been trapped in the pyrite grains. The formation of pyrite results from a slightly more oxidized environment, which also breaks up the sulfide complexes that are keeping the gold in solution. Gold nuggets are formed when gold precipitates in a crack or vug in a vein. The white quartz mentioned by the Australian prospector is a common gangue rock of hydrothermal veins, and its presence suggests that hot water has been circulating around in cracks in the rocks at some time. This hot groundwater often carries metals and precipitates them in vein deposits.

Once the gold has started to form, it can continue to grow until it fills the vug. Erosion then weathers out the gold, and since gold is unaffected by weathering, the gold nugget is formed. The same process gives rise to the tiny grains of gold one can find in river gravel in tourist gold-panning sites. In this case, the grains—called fines—have been eroded out of the rock and transported with the river sediment. The pyrite associated with the quartz oxidizes at the surface, leaving iron oxide as a rusty red stain. This can be seen most obviously when the host rock is white quartz, which, being hard, sticks out above the surface. This is the red-stained white quartz the prospector used as a guide to gold. So pyrite, which produces the red stain, can be an indicator of gold nuggets nearby. However, correlation in nature is not causation. Pyrite and gold are often found together because they are both related to a third factor: the change in the oxidation state of the solution as it nears the surface.

Another reason most people have never seen a real gold nugget is that many of the nuggets on public display in museums and up-market jewelers are fakes. Some years ago the Geological Museum in South Kensington, London, had a great collection of large gold nuggets on display as a part of its mineral exhibition—and also as the vestiges of the booty of empire. One day someone walked in and simply stole them. Since that time, museums have been more careful, and the gold nuggets you see now have been mostly replaced by replicas.

The driving force for this heist was the rapid increase in the value of gold. This rise in value was so rapid that the museum keepers had not twigged to just how much cash was now stored in their display cases. For years the world economy was based on something called the gold standard. This set the price of gold at US$35 per troy ounce, and the exchange rates of other major currencies were then fixed to this value. The troy ounce is the basic unit for weighing valuable metals like gold and is a bit more than the everyday ounce: 1 troy ounce equals 1.0971 avoirdupois ounces or 31 g, something you should remember while at the pawnbroker's. This fixed price meant that miners could receive no more than US$35 or its equivalent in another currency for every ounce of gold they produced. The gold standard ended in 1971 and the price of gold rose until it reached almost US$2,000 per ounce before falling back to around US$1,200 per ounce. The effect of this price rise on the mines was dramatic.[17] For example, if it costs US$100 to mine a ton of rock, that ton of rock must contain gold worth more than US$100 for the process to be profitable. When gold was US$35 an ounce, this would mean that the rock would have to contain around 3 ounces or 100 g of gold; when the price of gold is US$1,200 per ounce, the rock would need to contain only 2.5 g of gold per ton to meet the mining costs. Gold can be mined today from much poorer ores than in the past, and the small amounts of gold contained in pyrite become ore grade. Of course the period when mining gold was necessarily expensive was just a short period in human history. For much of historical time mining was free because it was carried out by slaves who enjoyed a very short lifetime. So any gold that could be recovered represented a profit.

The occurrence of invisible gold has given rise to a number of gold scams, so the wheel has turned full circle (Chapter 1). The most famous recent one was the Bre-X scam, which came to light when the mine geologist "fell out" of a helicopter. Bre-X owned a prospect at Busang in Indonesia, which the company claimed contained 70 million troy ounces of gold and was rumored to contain 200 million ounces. At a gold price of even US$1,200 per ounce, this is a lot of money. As with the Frobisher scam, this was a remote location (in the Borneo jungle in this case), and the original assay reports that produced the original high gold estimates were not reproduced by an independent assayist. Further study showed that the crushed samples of drill core that had been sent to the assay office had been salted with gold that had been shaved off gold jewelry. The value of the scam was at least US$3 billion, which is the amount investors lost in the collapse of the Bre-X shares. As a footnote, some people think that the jungle was also salted with a body: the mine geologist's body was identified by a molar and a thumbprint, and some people claim to have seen him in Canada after this event.[18] Busang is one of a number of gold prospects in this region of Borneo. The gold, of course, is invisible gold contained in pyrite. That old sea-rogue John Frobisher would have heartily approved.

Notes

1. *Homo sapiens sapiens* refers to the subspecies of *Homo sapiens* that includes anatomically modern humans.

2. A.E. Annels and D.E. Roberts. 1989. Turbidite-hosted gold mineralization at the Dolaucothi Gold Mines, Dyfed, Wales, United Kingdom. *Economic Geology*, 84:1293–1314.

3. M. Twain. 1883. *Life on the Mississippi* (Boston: James R. Osgood and Co.), 270pp.

4. Or *was collected*: the pyritic stockpiles at Kasese may have run out by the end of 2013.

5. C. Wadia, A.P. Alivisatos, and D.M. Kammen. 2009. Materials availability expands the opportunity for large-scale photovoltaics deployment. *Environmental Science and Technology*, 43:2072–2077.

6. Uncertain future for pyrite victims. 2012. *Northside People*, April 11.

7. B.Tuohy, N. Carroll, and M. Edger. 2012. *Report of the Pyrite Panel* (Dublin: Department of the Environment, Community and Local Government), 173 pp.

8. A.B. Hawkins. 2011. Sulphate-heave: A model to explain the rapid rise of ground-bearing floor-slabs. *Bulletin of Engineering Geology and the Environment*, 71:113–117.

9. A. Lugg and D. Probert. 1996. Mundic-type problems: A building mineral catastrophe. *Construction and Building Materials*, 10:467–474.

10. L.D. Bryant. 2003. *Geotechnical problems with pyritic rock and soil* (Master's thesis, Virginia Polytechnic University), 252pp.

11. D. Rickard, B. Hatton, D. Murphy, I. Butler, and A. Oldroyd. 2011. FeS-induced radical formation and its effect on plasmid DNA. *Aquatic Geochemistry*, 17:545–566.

12. M.J. Borda, A.R. Elsetinow, D.R. Strongin, and M.A. Schoonen. 2003. A mechanism for the production of hydroxyl radical at surface defect sites on pyrite. *Geochimica et Cosmochimica Acta*, 67:935–939. C.A. Cohn, M.J. Borda, and M.A. Schoonen. 2004. RNA decomposition by pyrite-induced radicals and possible role of lipids during the emergence of life. *Earth and Planetary Science Letters*, 225:271–278.

13. M.A.A. Schoonen, A.D. Harrington, R. Laffers, and D. Strongin. 2010. Role of hydrogen peroxide and hydroxyl radical in pyrite oxidation by molecular oxygen. *Geochimica et Cosmochimica Acta*, 74:4971–4987.

14. Y. Ma and C. Lin. 2013. Microbial oxidation of Fe^{2+} and pyrite exposed to flux of molecular H_2O_2 in acidic media. *Nature Scientific Reports*, 3: art. no. 1979.

15. See J.A. Imlay. 2003. Pathways of oxidative damage. *Annual Review of Microbiology*, 57:395–418.

16. See D. Rickard. 2013. *Sulfidic sediments and sedimentary rocks* (Amsterdam: Elsevier), 801pp. The reason pyrite is so ubiquitous on Earth is that it is the stable iron sulfide phase in water.

17. The effect of this price rise on national treasuries was even more dramatic. The hoards of gold that successive governments had stashed in Fort Knox and the Bank of England as the leverage to the currency values had been bought at US$35 an ounce and were sold off at fifty times that value. A nice little earner, as the cockney traders say in London.

18. For more about the Bre-X scam, see D. Gould. 1998. *The Bre-X Fraud* (Toronto: McClelland and Stewart), 272pp.

Epilogue

So we have come to the end of our great journey with pyrite. We have struggled through briar patches of esoteric science and waded the wetlands of cultural history; we have glimpsed the glorious sunlit mountain peaks of great intellectual achievement and visited dark regions of deep time when the Earth was young and easy. Our way in all of these excursions has been illuminated by pyrite.

We have seen that the fundamental reason for the preeminence of pyrite as a key material in the development of humankind and the evolution of the surface environment of the Earth itself is, as Archimedes would have recognized, that it has the properties of both stone and metal. Pyrite behaves like a common stone in its distribution in rocks of all ages and all regions of the planet but like a metal in the way it shines in the rocks. Its dual nature has meant that it has been a key material to humankind throughout the ages as a source of both valuable metals and sulfur. Its diverse properties have tested our philosophies throughout the ages: theories about how minerals form, the mechanics of Earth processes, and the fundamental constitution and basic attributes of matter have all been proven on the anvil of pyrite. Even today, this essential duality has been used in the electronics industry as it has been discovered that pyrite can be made to switch its dominant property between these two extreme states.

One of the best-received presents to children of all ages is a lump of bright golden pyrite. The delight people of all cultures, backgrounds, and education have in this mineral is quite astonishing—how they put it on display and hoard it as if it were a most treasured possession instead of one of the most common substances on Earth. It seems as though we are a simple species who, jackdaw-like, take delight in bright and shiny things. Certainly, New Age shamans in their magic crystal shops have been moved to associate the mineral with a vast array of unrelated properties. However, I cannot believe that the current response to pyrite is a recent human phenomenon; rather, it has all the atavistic characteristics of an ancient reaction. It would be interesting to see if our closest relations in the great apes react similarly—that is, regarding the crystals with evident delight rather than trying to eat them, for example.

Look around the room you are sitting in now and see if you can find anything that did not originate as a mineral. The more astute might point to the wooden table and forget the screws and nails that hold it together or the metal saws that cut the wood or the fertilizer that helped the trees grow. Others might point to the curtains and overlook the dyes that gave them their color or the original mordants of the cloth, let alone the scissors that cut them. Yet others would point to a plastic bowl, forgetting that plastics have their origins in petroleum. I have run this exercise with grade-school children, university students, and the general public, and I am always amazed about the lack of connection in the public consciousness today regarding the source of their material well-being and minerals. So it is not perhaps surprising that the role of pyrite in the development of our civilization has been widely overlooked. As far as I know, there are no history courses relating epic tales of pyrite rather than of cabbages and kings.

In some ways this book has sought to rectify this lacuna. I have woven tales around pyrite and demonstrated how this heroic mineral has molded the course of human development, shaped the natural environment of the planet, and provided insights into the way the world works.

Index

Italicized page numbers indicate a figure. Page numbers followed by n and a number indicate a reference to a numbered footnote on the designated page.